ELECTRONIC SYNTHESISER PROJECTS

ALSO BY THE SAME AUTHOR

BP51. Electronic Music and Creative Tape Recording

ELECTRONIC SYNTHESISER PROJECTS

by

M. K. BERRY

BERNARD BABANI (publishing) LTD
The Grampians
Shepherds Bush Road
London W6 7NF
England

First Published February 1981

British Library Cataloguing in Publication Data
Berry, M K
Electronic synthesiser projects.
1. Musical instruments, Electronic Construction
I. Title
789.9 ML 1092

ISBN 0 85934 056 2

Printed and Manufactuured in Great Britain by Hunt Barnard Printing Ltd.

CONTENTS

CHAPTER 1

INTRODUCTION

In the following pages are a number of electronic synthesiser related projects for a variety of sound effects. Each chapter deals with one project and begins by explaining its application and theory of operation. Most of the projects use integrated circuits, usually as the central element, but also in peripheral circuits. There are now a very large variety of I.C.'s on the market, which perform many tasks in the sound effects field at a much cheaper price than with discrete transistors. From the construction point of view much time can be saved and, with less soldered joints, there is a lot less to go wrong.

Full constructional details are given for all circuits. Printed circuit boards have been adopted as a general method of construction, but there is no reason why Veroboard or other methods cannot be used instead. It is assumed that most readers will be acquainted with the preparation of printed circuit boards. If not, full instructions are given with the many printed circuit kits that can be purchased. As regards cases and front panels, general suggestions only are given. Many readers may have existing equipment which they might wish to match into, or may have their own preferences with regards to housing the circuits.

Few aspects of the projects could be considered 'critical', and so construction of them could easily be undertaken by the inexperienced. Sockets can be used for the I.C.'s, thus reducing the possibility of damage while soldering them in. Setting up requirements are kept to a minimum, but where they are necessary a good multimeter is all you need. One that has a sensitivity of 20 kΩ/V would be adequate. An oscilloscope is an expensive luxury these days, but if you have access to one, it will be very useful for looking at the various waveforms. In addition, a simple amplifier with loudspeaker and an audio signal generator (sine/square wave output, 20 Hz – 20 kHz frequency) will be found most useful.

The final chapter deals with the projects as a whole, and how they may be fitted together. It also gives information on how the projects in this book can work within an electronic music studio, although it is expected that you, the reader, will have many more ideas on how these projects can be made to produce electronic music.

Finally, the author wishes to acknowledge the invaluable help of Miss Debra Berry for checking, correcting and typing the manuscript.

CHAPTER 2

SINGLE-CHIP SYNTHESISER

If you take a VCO, a low frequency oscillator, an envelope shaper and VCA, a noise generator and a mixer, plus various other bits and pieces, you have got a simple synthesiser. What may be surprising is that all of this is now available in a little black package with 28 pins protruding therefrom.

The SN76477N chip is now available for two or three pounds. It is primarily intended for use in games and alarms. However, due to the versatility of the device it is possible to program an extremely wide variety of audio waveforms.

Programming is by four methods, namely resistors, capacitors, logic and analogue signals, operating in various combinations.

Figure 2.1 shows a block diagram of the chip. Essentially there are three sound generators – the LFO, VCO and noise generator, which are mixed, given an envelope, amplified and finally outputted. Most parameters can be controlled externally and any combination of the three generators can be selected in the mixer.

We shall look at each block in the system in turn, concentrating on how it is controlled and what it controls after it.

Low Frequency Oscillator (LFO)

The LFO has a working frequency range from 0.1 Hz to 30 Hz, which is controlled by a resistor on pin 20 and a capacitor on pin 21. The frequency is given by:

$$f = 0.64/RC \text{ Hz}$$

The LFO produces two outputs – the first is a square wave which is fed to the mixer, and the second is a triangular wave which is fed to the VCO via a switching system. This second waveform can be changed to an exponential one by putting a resistor across the timing capacitor on pin 21.

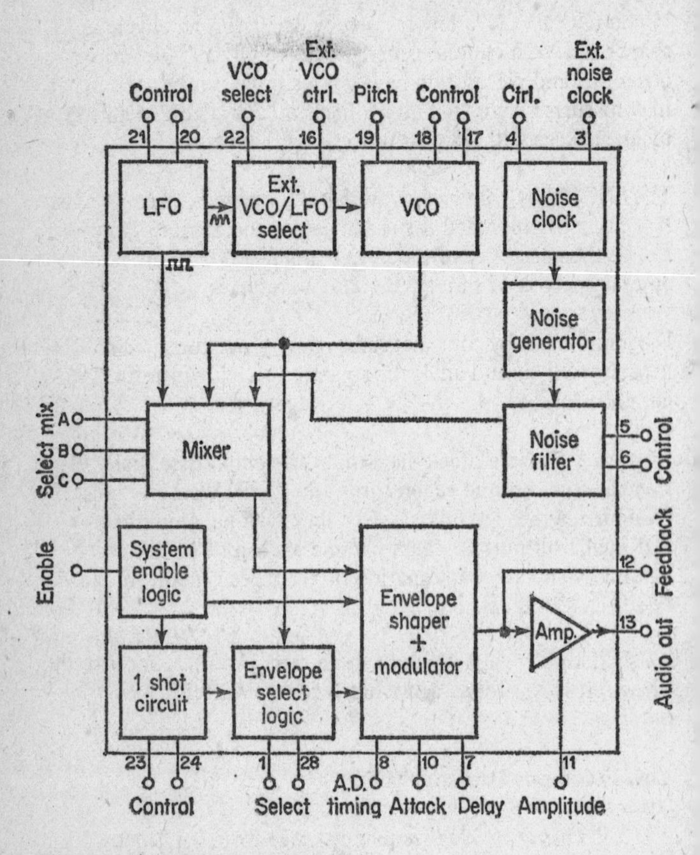

Fig. 2.1 Block diagram of 76477

Voltage Controlled Oscillator (VCO)

The VCO has a working frequency range from 1 Hz to 20 kHz, and this is controlled by

a) an external voltage (pin 16)

or

b) the triangular output from the LFO

and/or

c) a resistor and capacitor on pins 18 and 17 respectively.

In fact the resistor and capacitor set the minimum frequency of the VCO, and is given by:

$$f = 0.64/RC \text{ Hz}$$

Between the LFO and VCO is a switching circuit which selects either the LFO triangle output or an external control voltage as a control input to the VCO. The switching function is controlled by logic signals on pin 22. Therefore a logic 0 on pin 22 connects the VCO control input to an external control voltage on pin 16, while a logic 1 connects the LFO triangle output to the VCO input. The external control voltage must lie between 0 and 2.5V. Frequency increases with decreasing voltage.

A voltage on pin 19 adjusts pitch, or mark-space ratio. This facility permits a change in the quality and tone colour of the sound produced. This voltage should be between 0 and 2.5V., and the mark-space ratio increases with increasing voltage.

The output of the VCO, which is a square wave, is fed to the mixer and to the envelope select logic.

Noise Generator

Something that hi-fi buffs are forever trying to eliminate is that which puts a lot of life into electronic music. In the 76477, white noise is generated by a shift register, clocked by a ring oscillator (noise clock). It is then passed through a low-pass filter whose bandwidth can be varied externally by a capacitor at pin 6 and a resistor at pin 5. The cutoff frequency is given by:

$$f = 1.28/RC \text{ Hz}$$

It is possible to omit filtering by removing the capacitor and leaving a resistor at pin 5.

The noise clock's frequency can also be varied with an external resistor on pin 4. Alternatively an external clock can be connected via pin 3.

Mixer

The three outputs from the LFO, VCO and noise generator are fed to the mixer. This is not an analogue mixer in the normal sense because instead of summing the input signals it multiplexes them. No problems are caused for us however, since the mixer output is taken internally to the envelope shaper and modulator.

The contents of the mix are determined by the three control lines A, B, C on pins 26, 25, 27 respectively. Logic signals are presented to these lines according to the truth table in Figure 2.2 in order to achieve the respective mix required. Note the control situation with logic 1 on all inputs – the output from the mixer is inhibited.

C (27)	B (25)	A (26)	Output
0	0	0	VCO
0	0	1	LFO
0	1	0	Noise
0	1	1	VCO/noise
1	0	0	LFO/noise
1	0	1	LFO/VCO/noise
1	1	0	LFO/VCO
1	1	1	Inhibit (zero output)

Fig. 2.2 Mixer truth table

System Enable Logic/One Shot Circuit/Envelope Select Logic/Envelope Shaper and Modulator

The first three circuits are all basically logic circuits which provide control for the envelope shaper and modulator.

The system enable logic switches the signal output of the chip on or off depending on whether a logic 1 or 0 respectively are applied to pin 9. This circuit also operates in conjunction with the one-shot circuit, which is a monostable whose time constant is determined by a resitor and capacitor on pins 24 and 23 respectively. This time constant can be found from:

$$t = RC \text{ seconds}$$

The one-shot circuit enables many sounds like explosions or percussive effects to be made. The monostable is activated when pin 9 goes from logic 1 to logic 0. In practice this is easily carried out with a push-button arrangement.

The envelope select logic is a circuit which decides which signal will be fed to the envelope shaper and modulator. The selection is controlled by logic signals on pins 1 and 28, and the truth table for this is shown in Figure 2.3. This circuit works in conjunction with the envelope shaper and modulator, which generates and applies an attack-decay envelope to the selected input. How this works with the envelope select logic is also set out in Figure 2.3

Logic input pin 1	Logic input pin 28	Envelope selected	Attack ramp starts on....
0	0	VCO	Positive edge from VCO
0	1	Mixer only	System enable (pin 9) to logic 0
1	0	One-shot	System enable (pin 9) to logic 0
1	1	VCO with alternating cycles	Every other positive edge from VCO

Fig. 2.3 Envelope select table

The envelope shaper is a simple attack-decay type whose attack and decay functions are controlled by resistors on pins 10 and 7 respectively, and also a capacitor on pin 8. The resistors set the relative attack and decay times, while the capacitor sets the overall timing.

Output Amplifier

The envelope shaper and modulator feeds directly into the output amplifying stage. The gain can be varied externally, and the output voltage peak-to-peak will be 3.4 R_V/R_G where R_V is the resistor at pin 12 and R_G is the resistor at pin 11.

Since R_G at pin 11 controls the gain directly, it can be usefully employed for amplitude modulation. The recommended limits to its value are 22 kΩ to 220 kΩ.

There also exists, via pin 12, the possibility of providing an external input. This can be in the form of a voice signal, via a microphone and pre-amplifier, or an electric musical instrument such as a guitar or organ. Feeding such a signal in here has the effect of mixing the internal signal with external before being amplified and outputted.

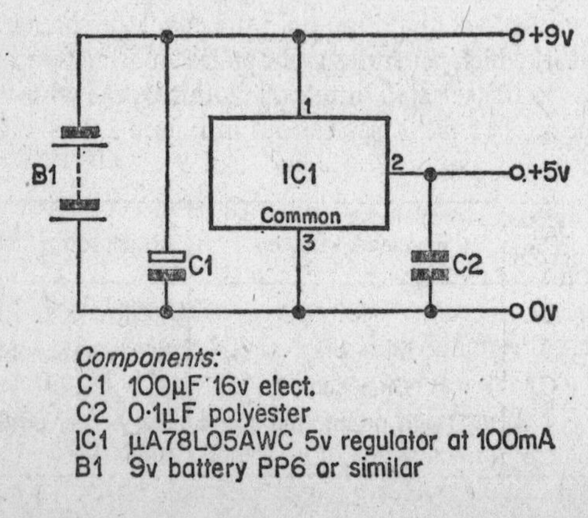

Fig. 2.4 Battery power supply

Regulator

This is the power supply of the chip. Externally, it requires a regulated 5 volts, an unregulated, but smoothed, 9 volts, and a ground. The current drawn by the chip is around 15 mA. This can easily be supplied by a battery, and a circuit to be used with one is shown in Figure 2.4. This circuit is required to give the regulated 5V. supply.

Figure 2.5 shows a mains power supply which will do the same job.

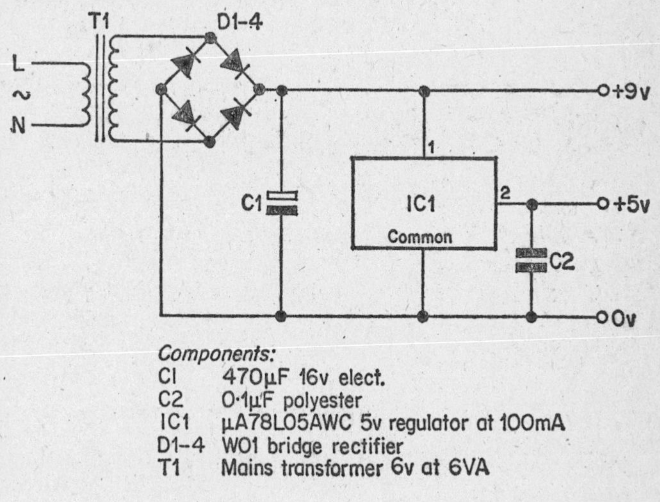

Fig. 2.5 Mains P.S.U.

A Practical Circuit

To enable the full potential of this chip to be realised, the circuit in Figure 2.6 has been designed so that all of the various functions of the chip can be varied. In most cases this means potentiometers or switched resistors/capacitors, but where logic signals are required simple on/off switches are used.

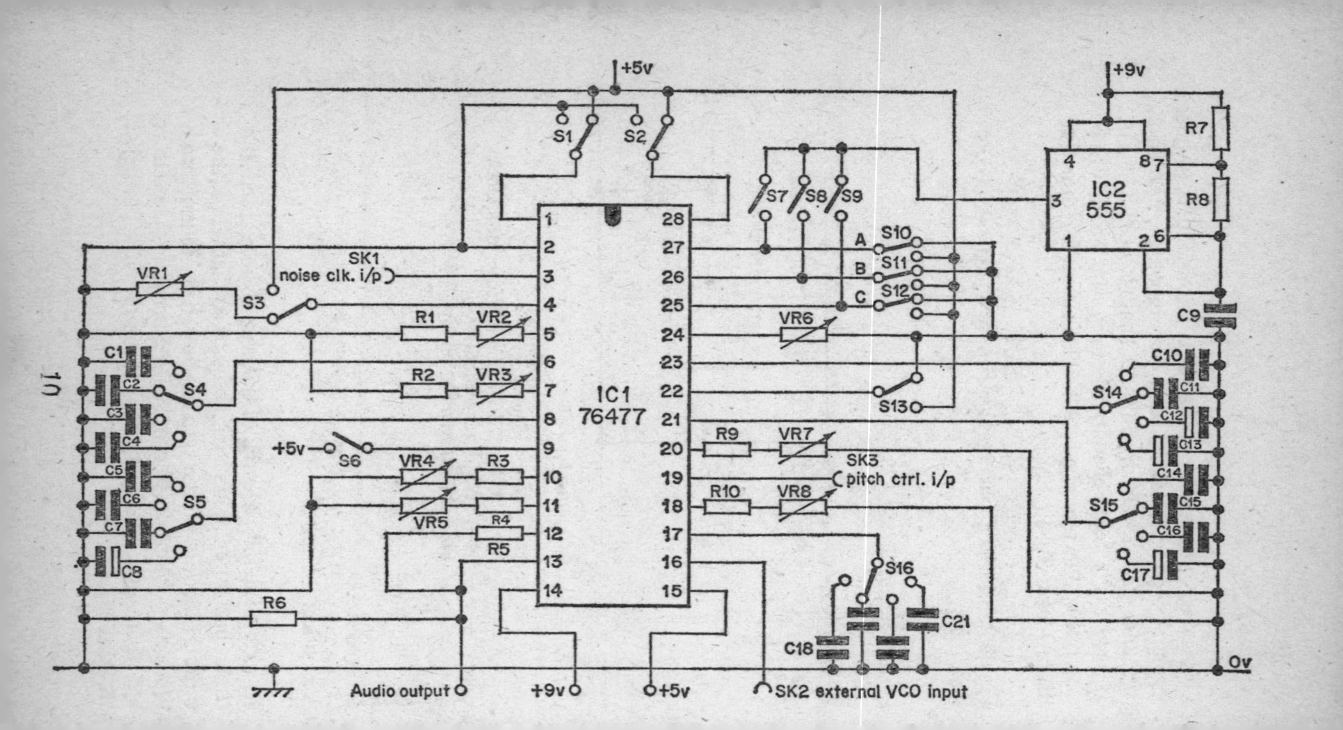
+5v
+9v
S1
S2
S3
S4
S5
S6
S7
S8
S9
S10
S11
S12
S13
S14
S15
S16
A
B
C
IC1
76477
IC2
555
SK1
noise clk. i/p
SK3
pitch ctrl. i/p
VR1
VR2
VR3
VR4
VR5
VR6
VR7
VR8
R1
R2
R3
R4
R5
R6
R7
R8
R9
R10
C1
C2
C3
C4
C5
C6
C7
C8
C9
C10
C11
C12
C13
C14
C15
C16
C17
C18
C21
0v
Audio output
+9v
+5v
SK2 external VCO input

Components:

R1	4·7k
R2	4·7k
R3	4·7k
R4	22k
R5	33k
R6	4·7k
R7	10k
R8	10k
R9	4·7k
R10	4·7k
all 1/4W	10% carbon
VR1	100k lin.
VR2–4	1M lin.
VR5	470k lin.
VR6–8	1M lin.
C1	330pF polystyrene
C2	1000pF polystyrene
C3	2200pF polystyrene
C4	0·01µF polyester
C5	0·01µF polyester
C6	0·1µF polyester
C7	1µF polyester
C8	10µF 12v elect.
C9	4700pF polystyrene
C10	0·1µF polyester
C11	1µF polyester
C12	10µF 12v elect.
C13	22µF 12v elect.
C14	470pF polystyrene
C15	4700pF polystyrene
C16	0·47µF polyester
C17	4·7µF 12v elect.
C18	100pF polystyrene
C19	0·01µF polyester
C20	0·1µF polyester
C21	1µF polyester
IC1	SN76477
IC2	NE555V
S1–3	1 pole changeover
S4–5	1 pole 4 way rotary
S6–9	1 pole on/off
S10–13	1 pole changeover
S14–16	1 pole 4 way rotary
SK1–3	single sockets to suit (e.g. phono, jack, etc.)

Fig. 2.6 Practical circuit for 76477

It would be easy to provide switches and variable controls for every possible patching arrangement, but some of these have been dispensed with in the interests of economy and simplicity of the final layout.

In this circuit, a clock generator has been provided which can be used to switch the mixer control lines A, B or C so that two simultaneous signals can be passed to the envelope shaper. This has the effect of providing the components of a mix with different levels with respect to one another. The rate of switching is about 60 kHz, which is well above the audible range.

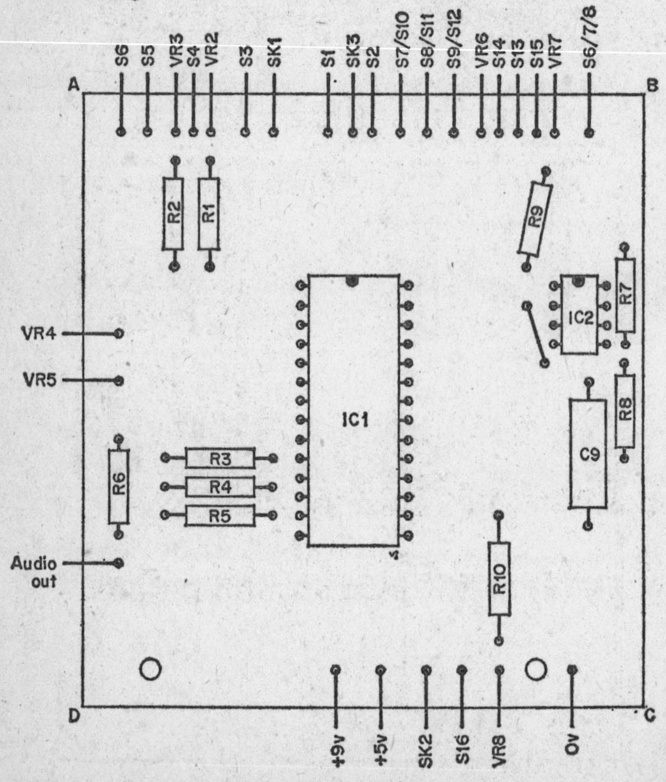

Fig. 2.7a Printed circuit layout for main circuit

There now follows full construction details.

Construction

The main chip and most of the associated components are mounted on a single printed circuit board. All potentiometers and switches are mounted on the case's front panel, connected to the p.c.b. by wiring. Figure 2.7 shows the layout and copper pattern for the main board, while Figures 2.8 and 2.9 show the boards for the battery p.s.u. and mains p.s.u. respectively. It is advisable to use a socket for the main chip – it will enable you to test the circuit before putting the I.C. in.

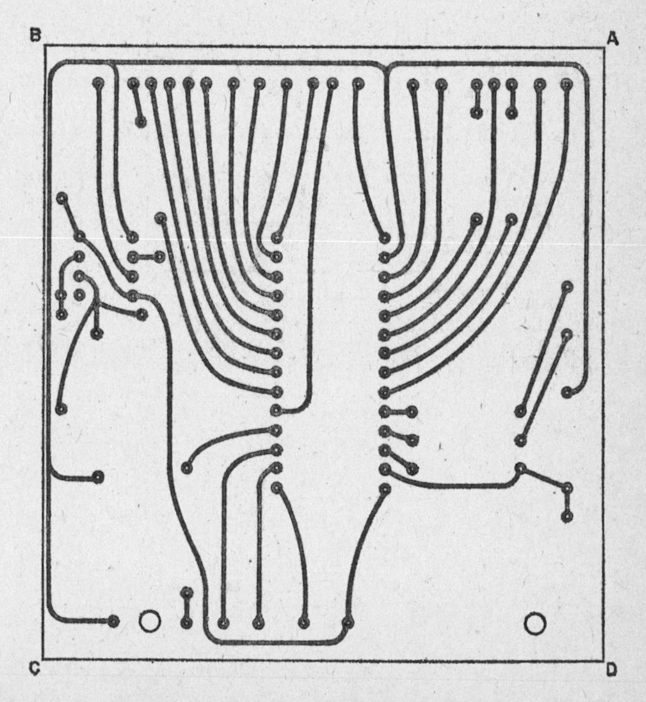

Fig. 2.7b Copper pattern for main circuit

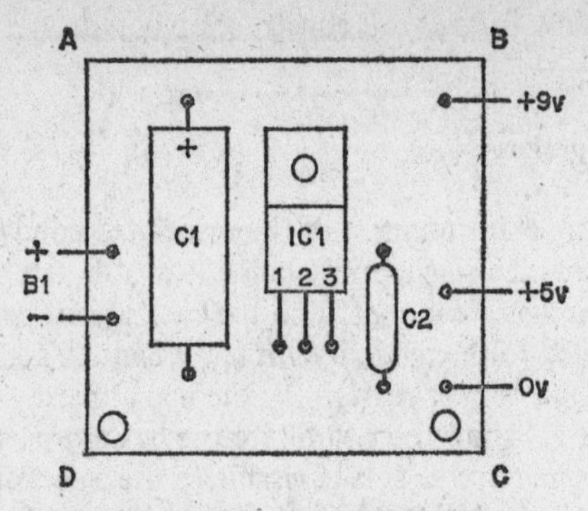

Fig. 2.8a Printed circuit layout for battery P.S.U.

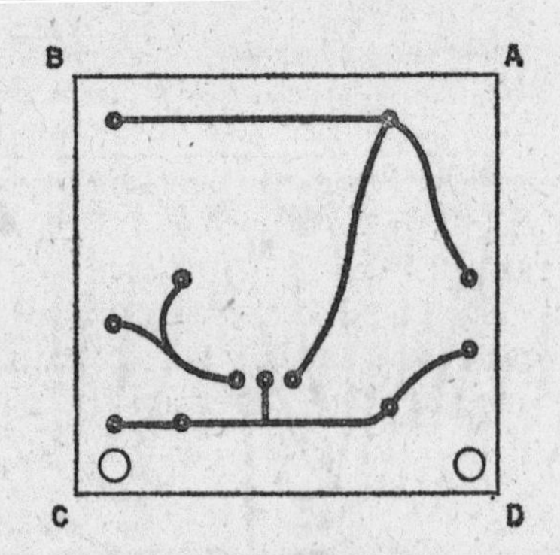

Fig. 2.8b Copper pattern for battery P.S.U.

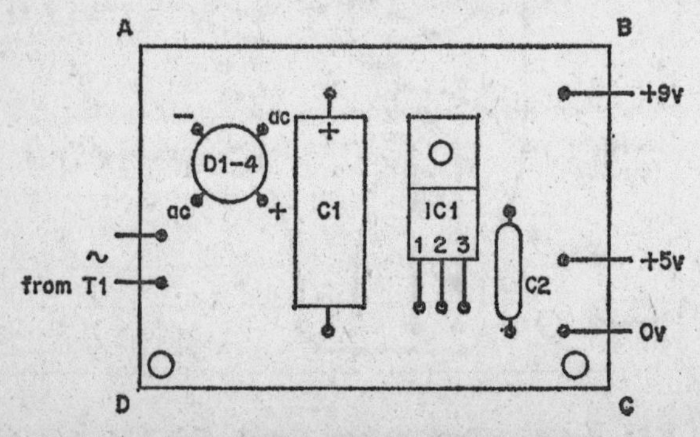

Fig. 2.9a Printed circuit layout for mains P.S.U.

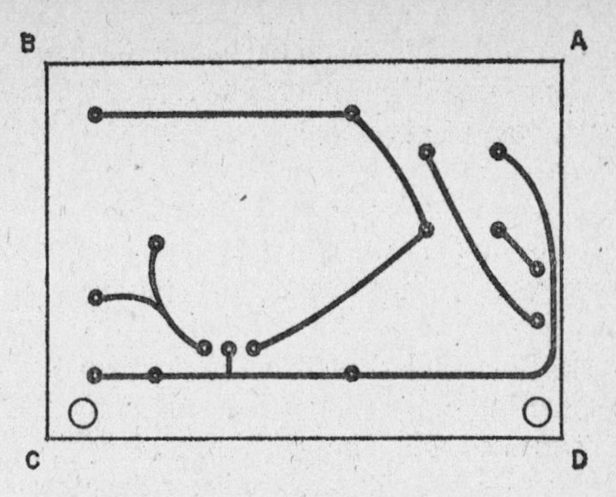

Fig. 2.9b Copper pattern for mains P.S.U.

The whole works can be housed in one case. The case should preferably be aluminium, to provide screening for the audio sections. One with a large removable front panel is ideal, because this panel will be able to carry all of the controls as well as the main board. This method of construction provides for a very neat layout and enables simple maintenance to be carried out. Figure 2.10 shows how the panel fits together. The sequence of construction is to build the main board first, and then mount the pots and switches onto the front panel. The panel, of course, should be pre-drilled, or punched, and lettered.

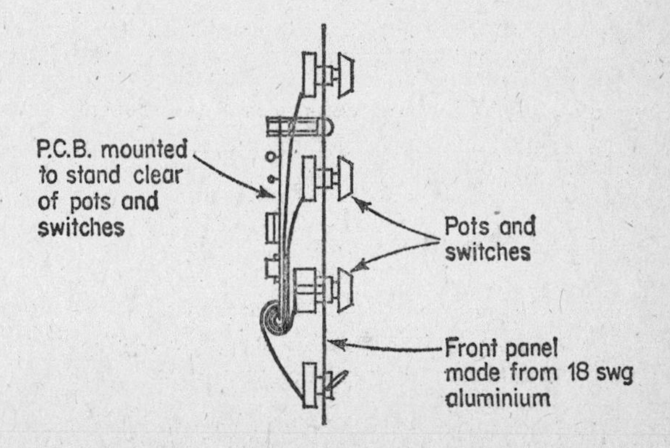

Fig. 2.10 Diagram showing method of panel assembly (not to scale)

Wiring can then begin, following the circuit diagram. The rule at this point is to keep all wiring to one side of the board, so that when the board fixing screws are removed, it can be 'hinged' away from the panel, exposing the controls. Use different coloured connecting wires if possible. Not only do they look professional but they make things a lot easier when you are trying to find out where the wires go!

When the panel is complete, you can start preparing the rest of the box. The power supply fits in here and fairly long wires should be connected to the board so that it is easy to lay the panel alongside the box without stretching the wires. The mains cable (if you are using a mains p.s.u.) is led into the box through a hole at the back. Make sure there is a grommet in this hole and it's a good idea to clamp the cable inside the box. Figure 2.11 gives details of the box assembly.

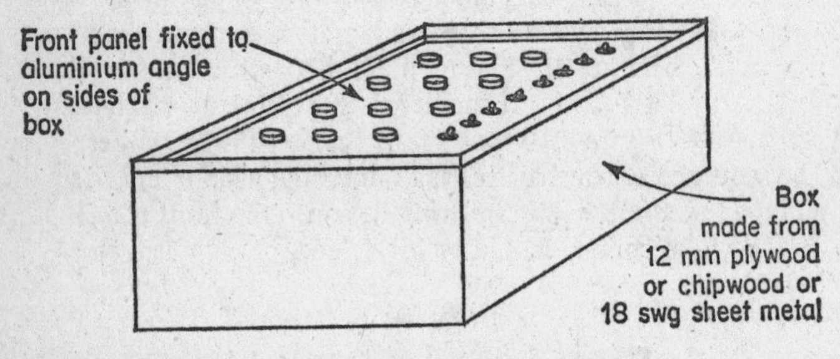

Fig. 2.11 Sketch showing panel in place in the box

After all this has been done, you can try the circuit out. Little should go wrong that a soldering iron will not cure.

Using the Single-Chip Synthesiser

After reading the explanatory notes earlier you should already have a good idea of what the 76477 is capable of, but here are just a few extra hints to help you on your way.

When setting up the circuit to produce a particular sound, it is best to adjust the controls in the following order:

1) Set all pots and rotary switches to their mid-values.

2) Select mixer combination.

3) Select envelope.

4) Adjust VCO, noise generator, and LFO.

5) Adjust attack and decay.

6) Adjust output amplitude.

A large variety of waveforms can be synthesised with this circuit, but further effects can be produced by bringing in the external circuits such as the clock on the mixer select lines or the the external noise clock. The VCO frequency also has provision for external control and this could be connected to a device such as the sequencer to be described later, or another voltage-generating device.

CHAPTER 3

ANALOGUE DELAY LINE

The concept of delay is very important in music and sound. Sound waves, travelling through air at 343 metres per second, take a finite time to reach the ear. Think of a gun being fired some distance away. You see the smoke, then hear the shot a couple of seconds later.

It is the delay in sound waves reaching our ears that enables us to position sound sources without using our eyes. Sitting in front of an orchestra, the sound waves emanating from a violin on the left take longer to reach the right ear than the left. A phase difference results, which the brain interprets as position.

Reverberation is also caused by delayed signals. More recent 'artificial' effects that rely on delay include phasing and flanging.

Until recently, these latter effects were either produced by two tape recorders playing the same signal – one delayed, one not – or by electronically filtering the signal so as to produce an effect approaching that of true phasing or flanging. Reverberation has been produced by all manner of mechanical methods: springs, plates, bathrooms and so on.

Now a purely electronic delay mechanism has been evolved – the analogue delay line, or bucket brigade device.

Theory

A most useful analogy is the bucket brigade, a line of firemen each with buckets. Water would be tipped from bucket to bucket along the line until it reached the fire at the other end.

In our delay line, the buckets are capacitors and the water is electric charge. The capacitors are connected in a chain, linked by electronic switches. The device requires a clock input to activate these switches. Figure 3.1 will help explain the method of operation.

On phase I of the clock signal, input 'a' is sampled and the capacitor in stage 1 charges up to the level of the signal. Phase II comes along and the contents of stage 1 are transferred to stage 2. No inputs are sampled on clock phase II. Phase I returns and another input signal 'c' is sampled and stored in stage 1. At the same time sample 'a' is shifted from stage 2 to 3.

It is not too difficult to see that each input sample gets shunted along the delay line every time the clock changes phase. Note, too, that an input is sampled only on alternate clock phases. To enable a complete signal to be taken from the far end of the delay line, the other half of the signal is clocked through another delay line which is connected in parallel to the first.

The clock input, however, must be in antiphase to that of the

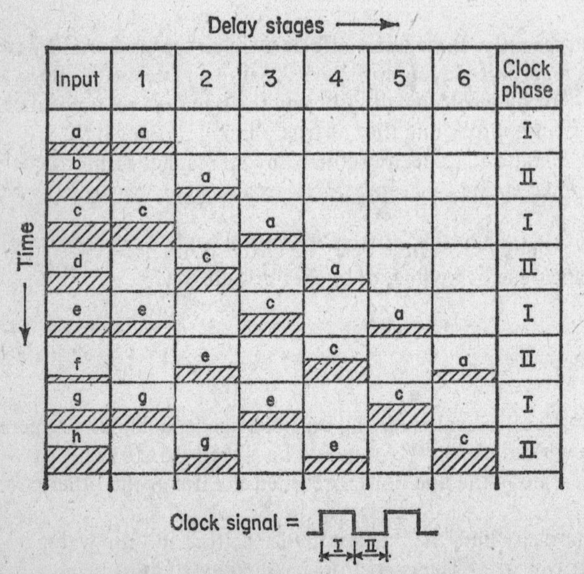

Fig. 3.1 Operation of delay line

first delay line, which is easily achieved in practice using digital circuit techniques, as will be seen later.

It is not too mathematically difficult to see that the delay time of a signal passing down the line is:

$$\frac{\text{number of stages}}{2 \text{ x clock frequency}} \text{ seconds}$$

So that if, as in the Mullard TDA1022, the number of stages is 512 (or rather 512 *pairs* of stages if you consider the two parallel lines), and the signal frequency is 1024 Hz, the delay time is:

$$\frac{512}{2 \text{ x } 1024} = 0.25 \text{ seconds}$$

It would seem that the analogue delay line wraps up the whole delay problem quite neatly in its black plastic box. Not so!

Of course, the trouble comes with bandwidth. As the clock frequency comes down and delay time goes up, so the bandwidth decreases in proportion. The maximum theoretical signal frequency is half the clock frequency, but practically it is better to assume the maximum bandwidth (i.e: maximum signal frequency) to be a third of the clock frequency.

To illustrate the effect which this limitation has on our use of the delay line, Figure 3.2 shows graphically how the bandwidth decreases with increase in delay time for the TDA1022.

If we wanted a delay of 0.1 seconds, the clock frequency would have to be:

$$\frac{512}{2 \text{ x } 0.1} = 2560 \text{ Hz}$$

Thus the maximum bandwidth would be:

$$1/3 \text{ x } 2560 = 853 \text{ Hz}$$

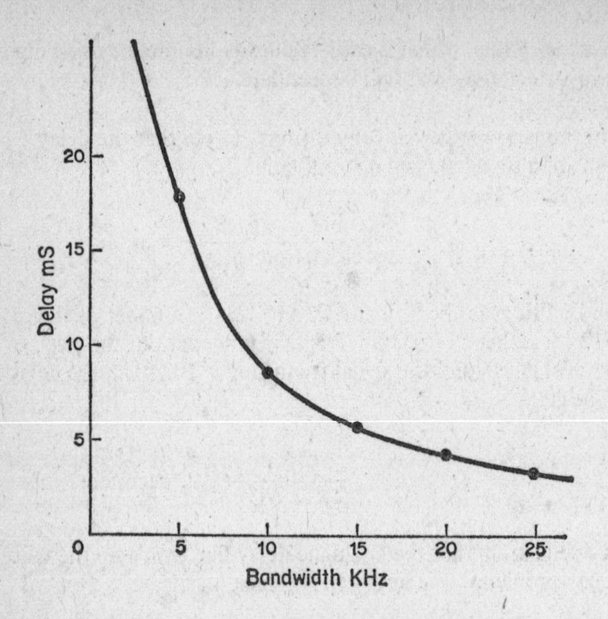

Fig. 3.2 Graph of delay vs. bandwidth for the TDA1022

It is easy to see that to achieve a workable bandwidth, we have to accept an apparently small delay time, unless we connected several delay lines together, which would be expensive.

In actual fact, a great many applications of delay lines only require a short delay time, and thus the bandwidth problem only has an effect when we want discrete echoes.

To prepare the signal for input it is passed through a low-pass filter to remove all frequencies above the bandwidth. In addition a low-pass filter is used at the output to remove the clock frequency.

Noise is always a problem in audio circuits, and the delay line is no exception in this respect. The noise arises from the leakage from the buckets, and is unfortunately more prominent at low frequencies. By using some form of noise reduction system it is possible to get signal to noise ratios of 70 dB.

Other problems include susceptibility to overload, requiring limiting on the input. Distortion due to the delay line is around 1%.

So much for the theory and associated problems. Now we will see what it can do in practice.

Basic Delay Line Circuit

As stated before we need to drive the TDA1022 with a two phase clock. Figure 3.3 shows a suitable circuit. It uses an NE566 function generator. It has a manual frequency control VR1 and also provision for voltage control via C1. This facility will be very useful when we come to the various applications of the delay line. Tr1 adjusts the DC level of the clock signal before it enters the flip-flop IC2 which produces the two complementary phases that are required by the delay line.

The next part of the basic circuit is the low-pass filter. In fact two of them should be constructed, one for the input to the line and one for the output. The filter is of a type known as a fourth-order Butterworth. As can be seen in the diagram (Figure 3.4) it consists of two similar stages. Each stage has a section which can vary the cut-off frequency from 500 Hz up to 15 kHz. The whole filter has a roll-off of 24 dB. for every doubling of frequency after the cut-off point.

Having dispensed with the essential peripherals, we can now deal with the delay line itself. Figure 3.5 shows the circuit detail. For what it does it is a surprisingly simple circuit – this is a reflection on how micro-electronics have changed practical construction of complicated circuits.

The two preset resistors in the circuit are for:

a) setting the DC bias at the input, so that the signal is symmetrical about the DC level.

b) balancing the outputs from the two delay lines so that there is a minimum breakthrough of the clock frequency.

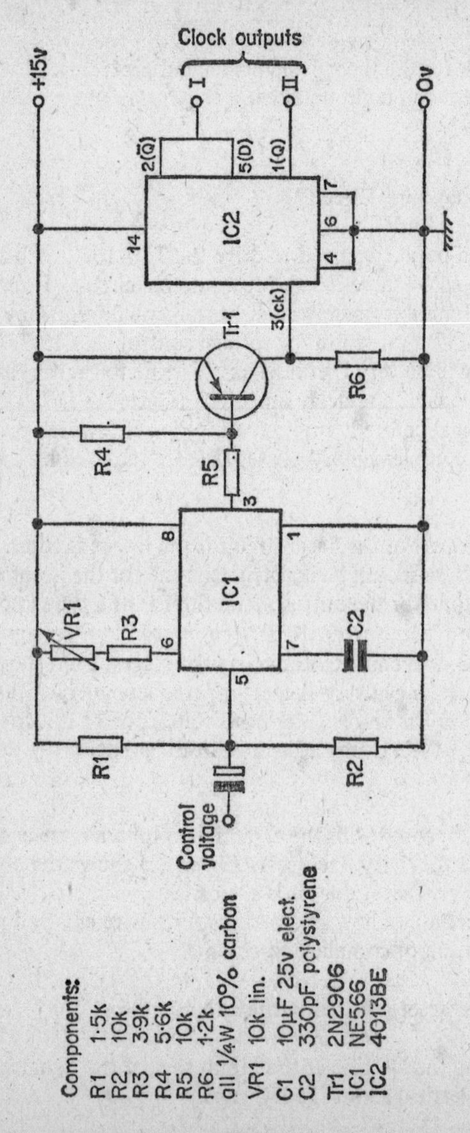

Fig. 3.3 Clock generator for TDA1022

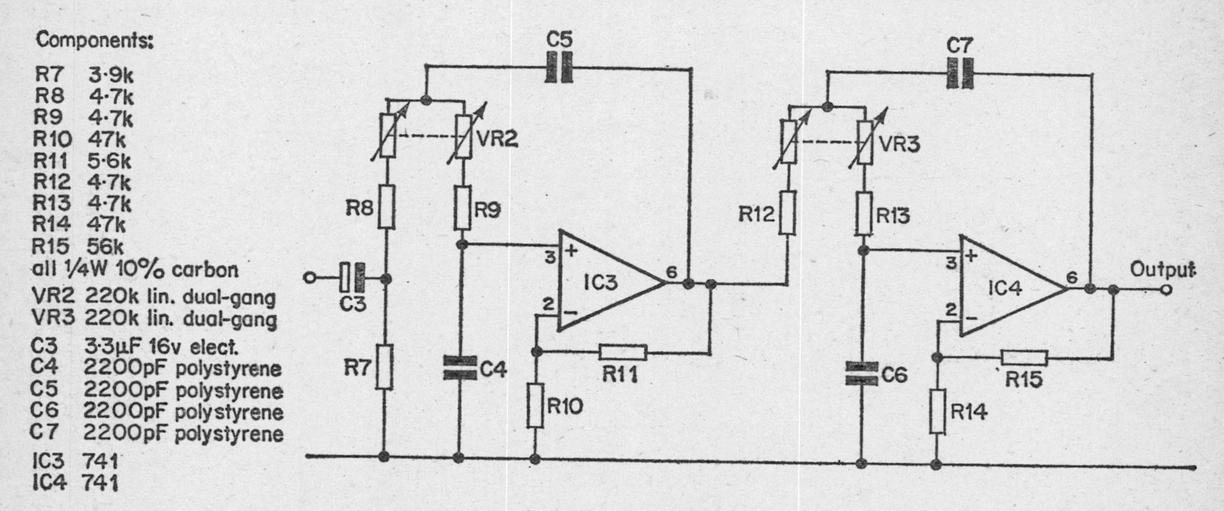

Fig. 3.4 Circuit diagram of low-pass filter (2 required)

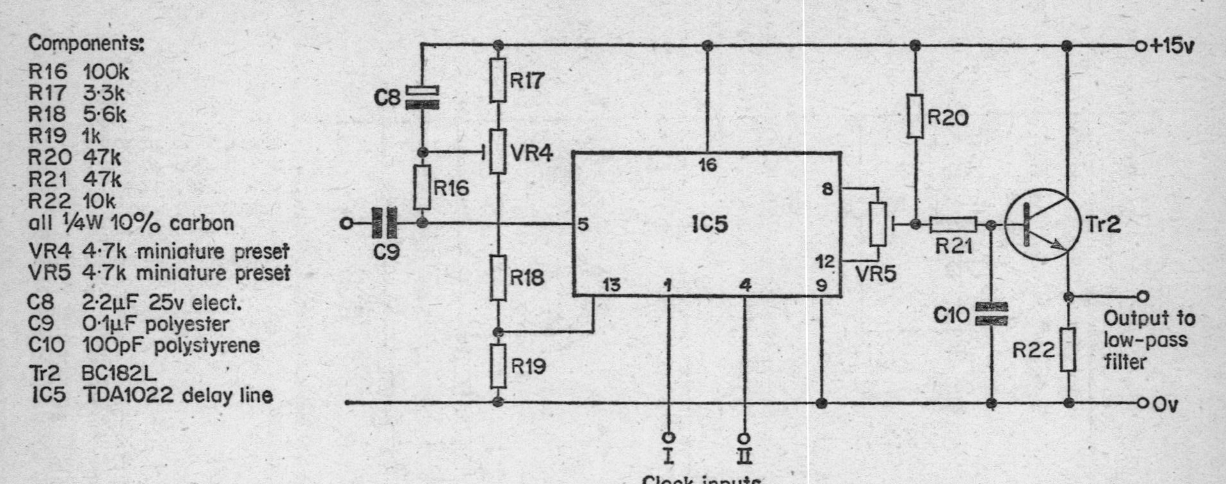

Fig. 3.5 Delay line circuit

Power Supply

The delay line and clock generator circuits require a +15 V supply, while the filters need a ±15 V supply. Thus a ±15 V supply is sufficient, and will also be able to supply power to additional circuits which we may wish to add in due course.

A mains power supply unit is shown in Figure 3.6. It is fairly straightforward, and uses a single chip to obtain regulated ±15 V.

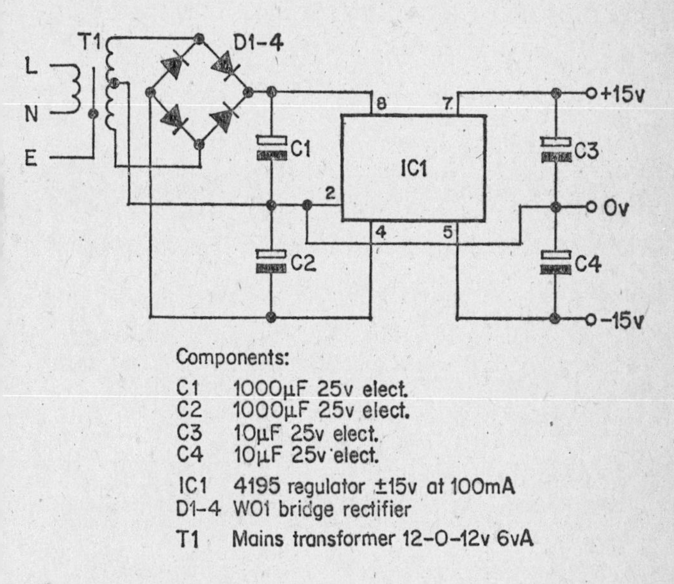

Fig. 3.6 Mains powered P.S.U. circuit diagram

Construction

The printed circuit layout and copper pattern are given in Figures 3.7 and 3.8 respectively. Included on this board are the delay circuit, its clock generator and two low-pass filters. The power supply is mounted on a separate p.c.b. which is shown in Figures 3.9. In fact the whole p.s.u. can be constructed

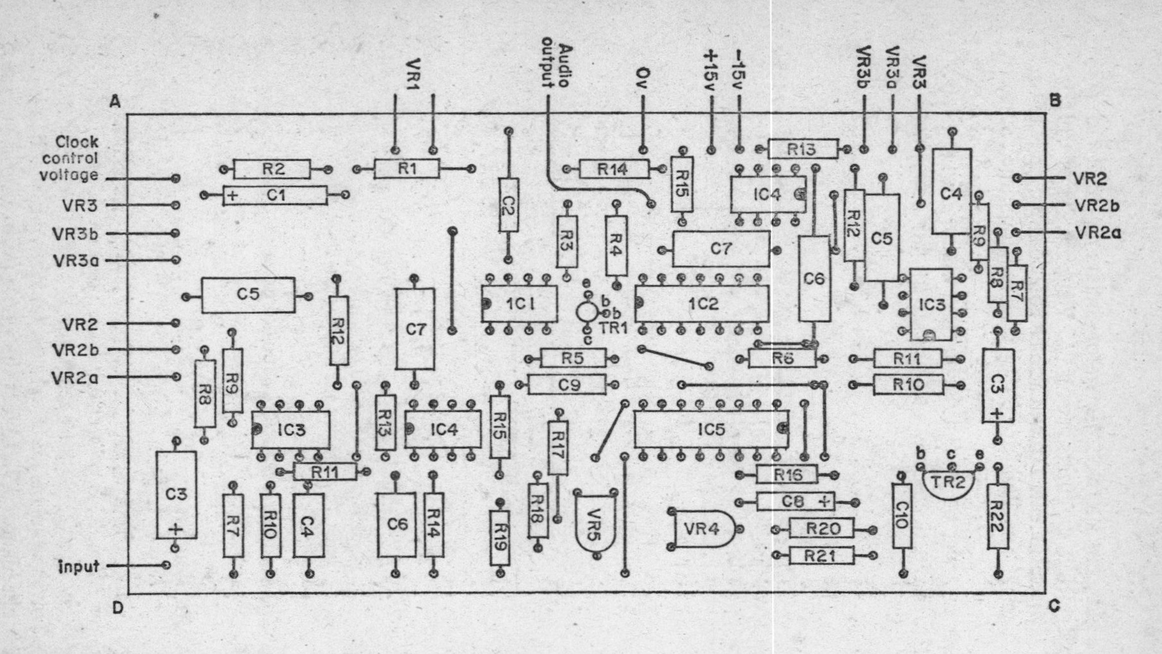

Fig. 3.7 Printed circuit layout for delay line circuit

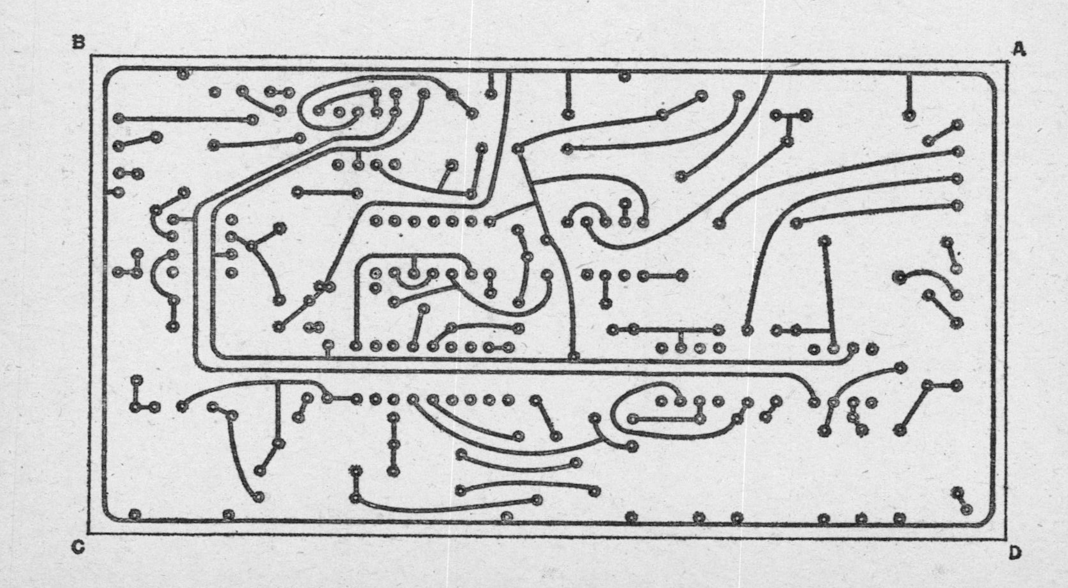

Fig. 3.8 Copper pattern for delay line circuit

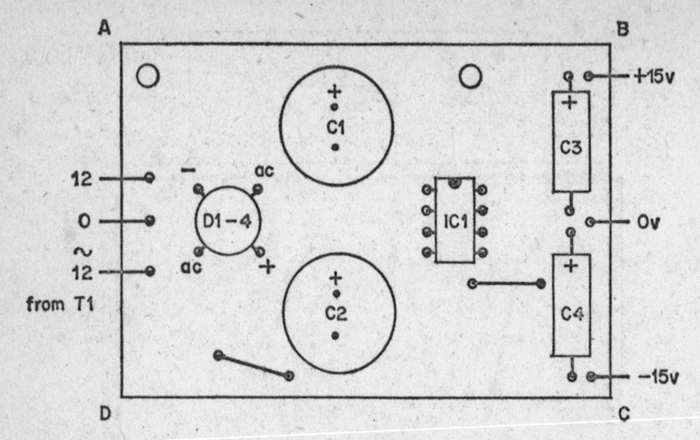

Fig. 3.9a Printed circuit layout for P.S.U.

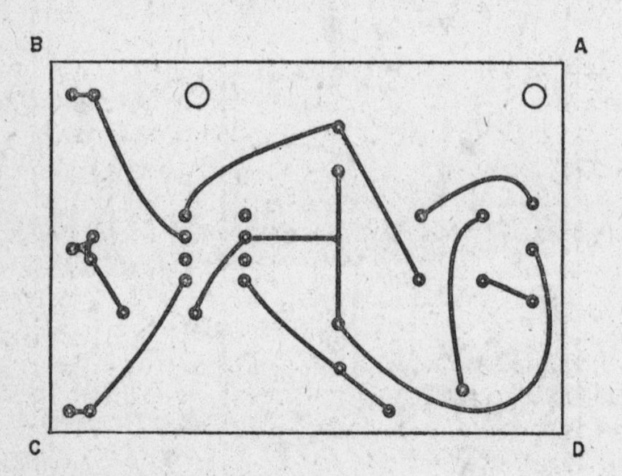

Fig. 3.9b Copper pattern for P.S.U.

as a complete module on a piece of aluminium sheet, such that other circuits are shielded from any hum interference from the transformer. The module can then be installed inside a case with other p.c.b.'s. Figure 3.10 shows how the p.s.u. module is made up. A terminal block is provided for the low voltage outputs so that more than one circuit can easily be connected. Details of housing the delay line circuit board will be given later, since there will be more boards to go in the same case.

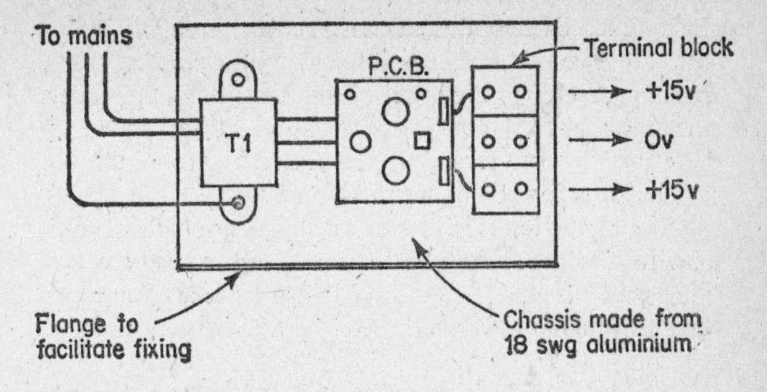

Fig. 3.10 Construction details of P.S.U. module

Setting-up the Delay Line Circuits

It is not a bad idea to construct the board in sections, so that each section can be tested separately, before going onto the next. The order of construction is given below, together with any testing required.

1) **Clock generator.** A simple check to see if the circuit is oscillating is to connect each output in turn to the line input of an amplifier so that the clock can be heard to oscillate when the frequency is set to its lower end. Turning the frequency control will increase the pitch until it soon goes out of audible hearing range. To check that the two outputs are in antiphase, just connect both to the amplifier input. The level should drop sharply indicating that the two signals are cancelling one another out. Of course, if you have an oscilloscope, it is much easier to check the operation of the clock.

2) **Low pass filters.** These can be checked by connecting an audio signal generator to the input and an amp. and speaker to the output. Each cut-off frequency can thus be calibrated by listening for the point where the level begins to fall off or by observing this point on a VU meter. Note that the signal generator must of course have the same output level for all frequencies.

3) **Delay line.** Once you have constructed this circuit, try out the whole board. With everything working connect the signal generator to the input of the first filter and connect the amplifier to the output of the second filter. Now increase the input level until the output begins to clip – a harsh distortion will be heard. Now adjust VR2 until the level of distortion is as low as possible – you will not be able to eliminate it entirely. Turn the input level down to a reasonable level, and adjust VR3 until there is minimum breakthrough of the clock frequency.

The delay line board is now set up and ready to use.

We will now look at the various applications of the delay line, beginning with phasing.

Phasing

This effect has in the past been produced by using a comb filter whose frequency notches were continuously varied. The delay line can become a comb filter by mixing the original signal with the delayed signal. As the two signals are a few milliseconds apart, there will be some cancellation at frequencies spaced 1/t Hz. apart, where t is the delay. By varying the delay time, therefore, a phasing effect can be produced. Figure 3.11 shows a block diagram of the set-up required.

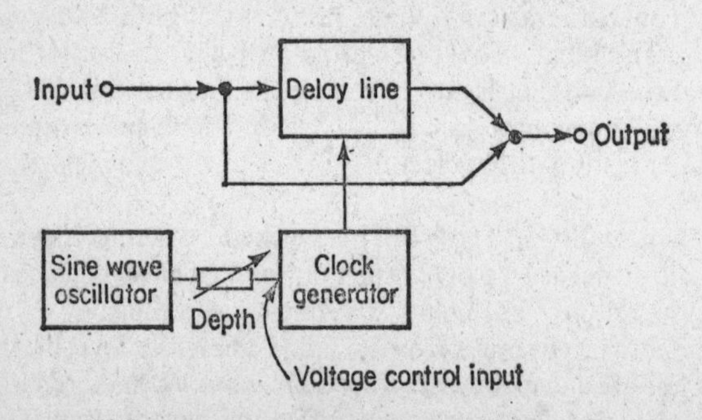

Fig. 3.11 Block diagram of phasing arrangement

A sine wave oscillator is used to modulate the clock generator's frequency at a slow rate of about 0.5 Hz. The output of this oscillator is fed to the voltage control input of the clock generator. Figure 3.12 shows a circuit diagram of the oscillator and Figure 3.13 gives board details and layout. Note that there is provision in the circuit for the frequency to be varied. Although we do not really need this for phasing, it will come in useful for other effects.

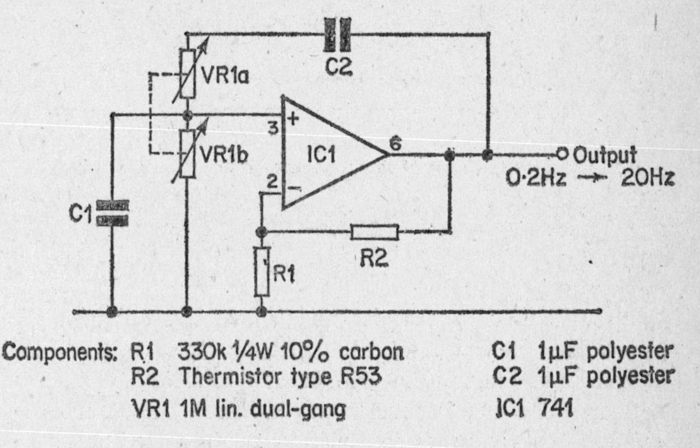

Fig. 3.12 Circuit diagram of sine wave oscillator

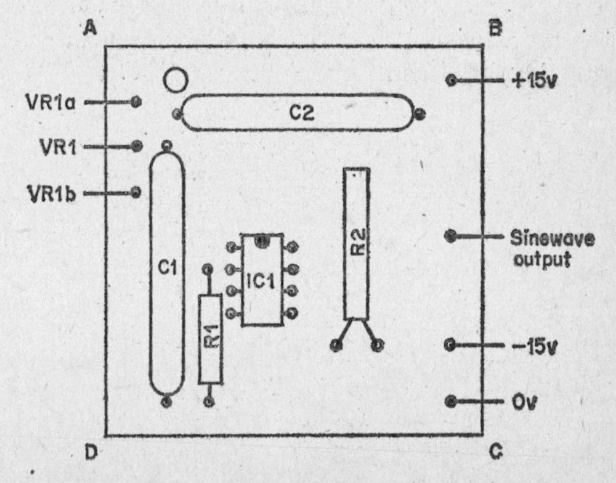

Fig. 3.13a Printed circuit layout of sine wave oscillator

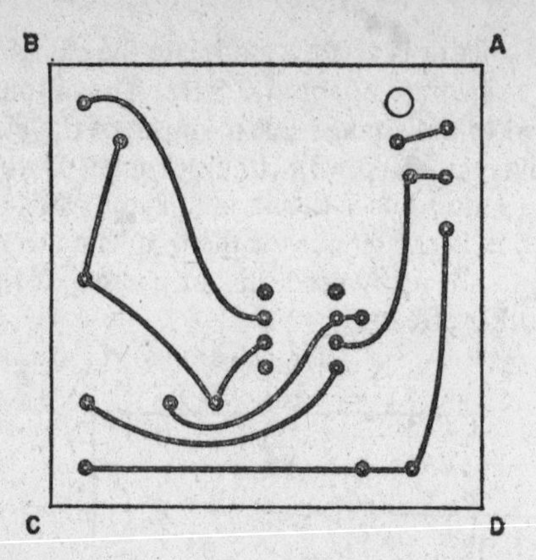

Fig. 3.13b Copper pattern of sine wave oscillator

Flanging

This effect is similar to that described above. It is obtained in much the same way, except that instead of mixing input with output you mix output with input – sounds confusing! What happens is that feedback is provided from the output so that, instead of frequency cancellations taking place, the response peaks at various frequencies. This gives a considerably coloured output. Figure 3.14 gives the details of how the effect is achieved. It uses virtually the same components as for the phasing effect.

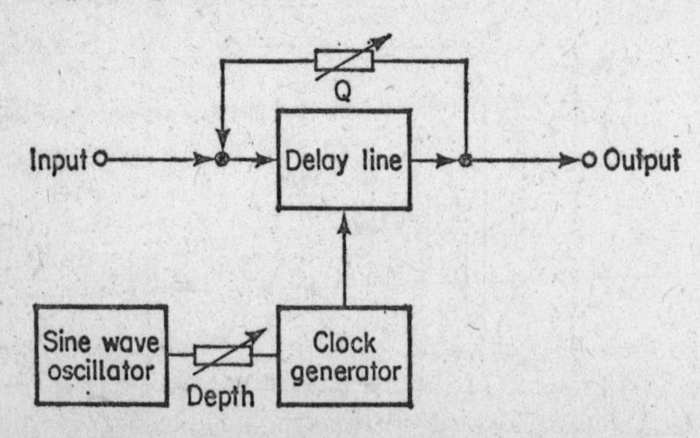

Fig. 3.14 Diagram of flanging arrangement

Vibrato

True vibrato is defined as frequency modulation, and before the delay line arrived it could only be achieved at the signal source. A vibrato effect could be achieved with filters and oscillators but, nevertheless, it was not the real thing.

In Figure 3.15 a block diagram shows how the delay line is used to create the effect. By modulating the clock generator with a slow sine wave the delay time can be varied. If a constant frequency is fed into the delay line the output will have a true vibrato with the rate being that of the sine wave oscillator. Once again the sine wave oscillator of Figure 3.12 is used to vary the vibrato rate.

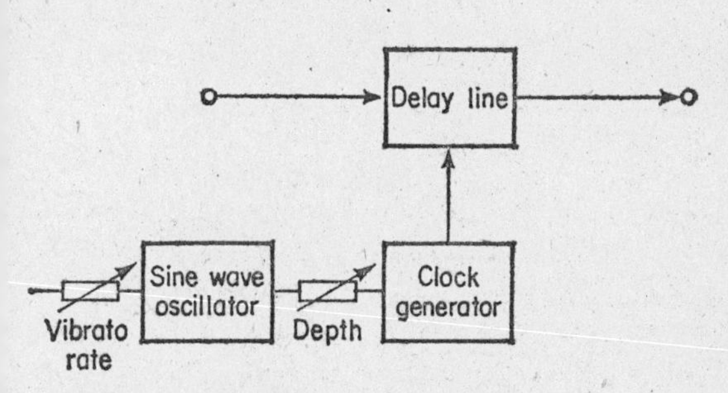

Fig. 3.15 Block diagram of vibrato arrangement

Echo and Reverberation

These two effects are basically the same, the difference being in the delay time required. For echo you need to have a long enough delay time so that each echo is a discrete sound, whereas for reverb the delay time is much shorter so that the several echoes merge into one, and the effect is of the sound dying away over about 1 second. Until now many methods have been used to generate these effects. For echo the most popular method is to use a tape loop with 5 or 6 heads around its circumference. Reverb methods include mechanical springs and plates with transducers attached to them and also using a special hard-walled room with a microphone at one end and a loudspeaker at the other.

With an analogue delay line things become much simpler. The method is to form a loop such that the delayed output is fed to the input – at a lower level than its original input level. A diagram of this arrangement is shown in Figure 3.16. The clock frequency, and therefore the delay time, can be varied by the potentiometer in the clock generator circuit. This controls whether you produce a short duration reverb or several discrete echoes. With one delay line you are not going to get very long echoes, but if these are required it is possible to connect more delay lines in series. Potentiometers are shown in the diagram which control the final echo/reverb output level and the level of the feedback. The latter control determines the rate of decay of the echo or reverb.

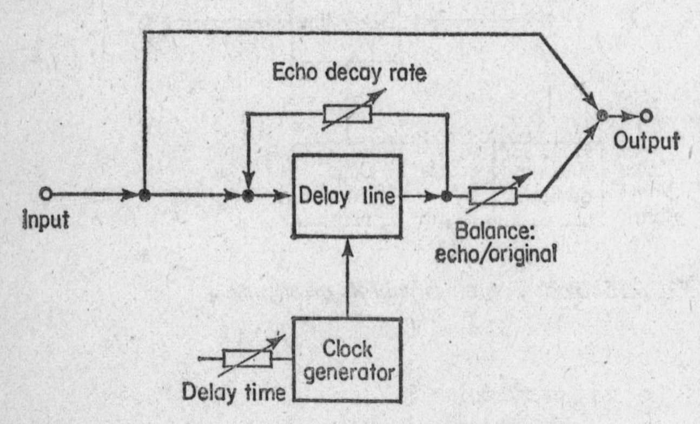

Fig. 3.16 Block diagram of echo/reverb arrangement

Double tracking is a particular use of echo which can give the impression that a second instrument is playing. This is done by adjusting the echo circuit so that a single echo is heard, delayed by about 50 mS.

From double tracking we move naturally on to chorussing. This is the effect which leads us to believe that there are a number of instruments playing, when in actual fact there is only one. It is used in the so-called 'string synthesisers', and can be a very pleasing sound. The effect requires more than one delay line – at least three preferably. Each delay line has its own slow sine wave oscillator as shown in Figure 3.17, and the delayed outputs are mixed together.

All of these effects have been achieved with only a few extra components. The main auxiliary circuit is the slow sine wave

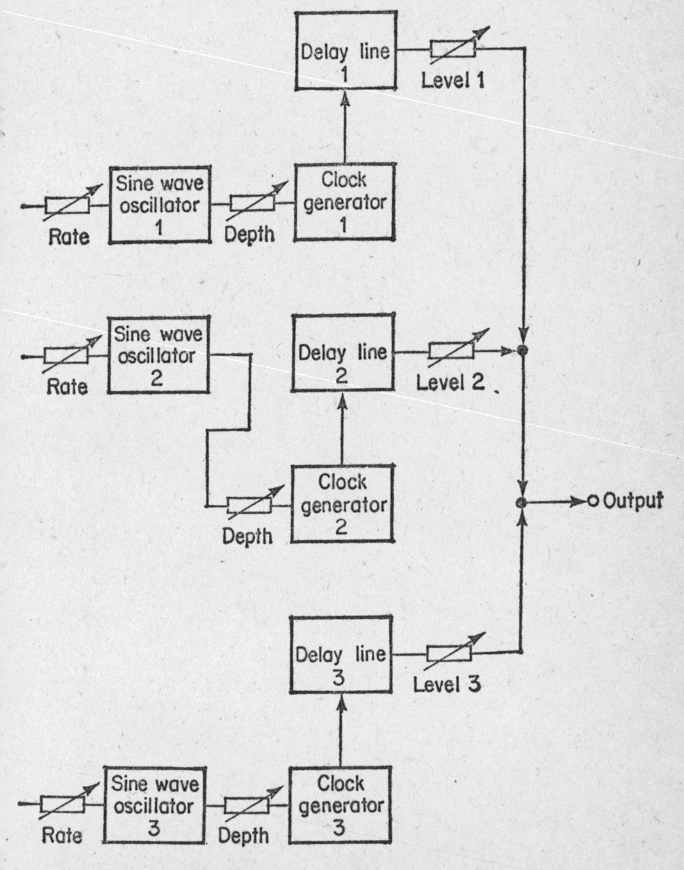

Fig. 3.17 Block diagram of chorus arrangement

oscillator. This can be housed in the same case as the delay line itself. To enable all the effects described earlier to be produced, several switches and pots need to be incorporated. Precise details are not given here since most readers will have their own preferences as regards layout and patching arrangements, and which of the effects they will want to use.

CHAPTER 4

PROGRAMMABLE SEQUENCER

The average electronic synthesiser consists of a number of different circuits, often duplicated, which can be linked in a variety of ways to enable different sounds to be synthesised. Most of these circuits are voltage-controlled, which means that their important parameters (frequency, amptitude, filter pass range etc.) are varied by means of a control voltage input. Usually the system is so designed that any circuit's output can be used to control another circuit via its control voltage input, and vice versa. In this way the versatility of the instrument is increased to a great extent. Having established that voltage control is the design principle that runs a synthesiser, we can see that a means by which a sequence of voltages can be memorised would be extremely useful. Such a piece of equipment is known as a sequencer. It is used usually in conjunction with – though not necessarily – a VCO, so that a tune can be programmed and played back at will.

There are basically two types of sequencer, digital and analogue, described below.

1) **Digital sequencer.** This usually has a digital memory chip (RAM) at its centre. To program it, a keyboard is used to write in each note as a binary number. The keyboard has a set of contacts which convert the number of the key pressed into this binary number. As each note is written into the memory, it is assigned an address and then the memory steps on, ready to receive the next note.

   For playback, a variable-speed clock is used to step through the memory, releasing each binary-encoded note, and sending it to a digital-analogue converter, where it is converted into a voltage, used to control a VCO.

   The advantage of this type of sequencer is that the memory used is capable of remembering a large number of notes –

256 is not uncommon. However, the pitch of each note has to be programmed beforehand, and cannot be set while the sequencer is running.

2) **Analogue sequencer.** The operation of this type of sequencer centres around a decade counter. This is a chip, or piece of logic circuitry, that sends a logic 1 to each of 10 outputs in turn. The outputs are connected to an electronic switch which switches in the output voltage from a potentiometer so that this voltage can then be used to control a VCO. A clock is used to switch the decade counter from one potentiometer to the next so that a sequence of voltages is sent to the VCO.

   With this type of sequencer, it is thus possible to adjust the pitch of the notes while the sequencer is running,– what could be called a real-time machine. The main disadvantage is that, because every note has a separate potentiometer, the number of notes in a sequence has to be limited to, say, twenty. If the main use of the sequencer is to generate background rhythms then this should not be a problem.

The sequencer to be described here is of the analogue type. Figure 4.1 shows a block diagram of the system. At first sight it probably looks extremely complex. In fact, it is a fairly simple piece of equipment that can produce complex sounds. Let us look at the diagram step by step.

At the front end is a clock with a variable frequency between 0.5 Hz and 20 Hz. This provides a pulse output which sets the rate at which the sequencer runs when playing. As an alternative to the clock is a one-step switch, which, as its name implies, steps the sequencer one step at a time. This facility will be particularly useful when programming the sequencer.

After the clock or one-step the pulses are split into two channels. In each channel the first block is the 'Divide by 1–10' unit. In fact it divides by 2–10, with an extra switch

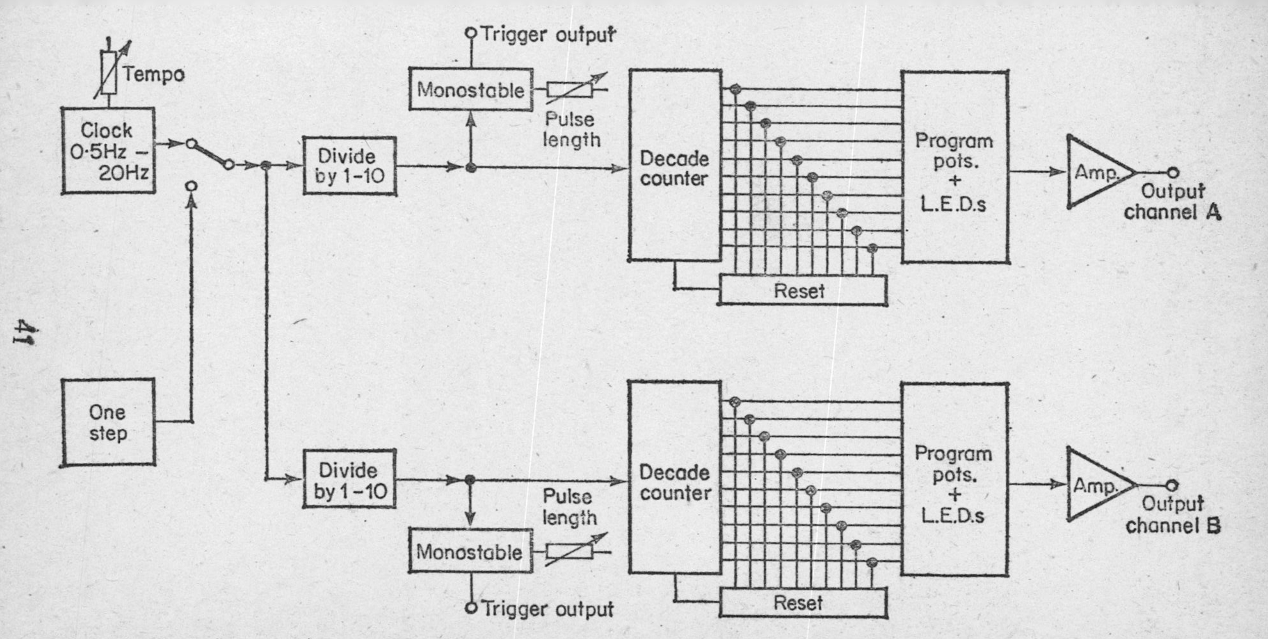

Fig. 4.1 Block diagram of sequencer

which allows the input pulse to by-pass the divider. The reason for these dividers is so that each channel can run at different rates, but where these rates bear a pre-programmed mathematical relation to one another, and are also synchronised. It may be difficult to see the point of this at present, but when we look at how to use the sequencer it will become obvious.

The dividers are followed by the decade counters, which splits the pulses up, sending them to each of 10 outputs one after the other. By connecting any of these outputs to the 'reset' pin of the counter, the sequence length can be shortened.

Each of the counter's outputs, on producing a pulse, causes a potentiometer to be connected to the main output. An l.e.d. is also connected to each of the counter's outputs so that it is easy to see which pot is active at any one point during the sequence.

An amplifier takes the output voltages from the pots and amplifies them to a suitable level to drive a VCO.

Circuit Operation

The clock generator and one-step switch circuits are shown in Figure 4.2. The clock is a simple CMOS astable, where VR1 sets the rate of oscillation between 50 mS and 1.5 s. IC1 is a set of four 2-input NOR gates. Two of them we used in the clock as inverters. The other two are used in the one-step circuit – one as an inverter, the other as a NOR gate. The purpose of the one-step circuit is to provide a single pulse to the decade counters. A simple push-button wired to the power rail would be fine, except for the possibility of switch bounce. The circuit eliminates this by employing a monostable to hold the pulse on for a ¼ second.

These two circuits are followed by two identical channels, A and B. The first two circuits in each channel are the divider

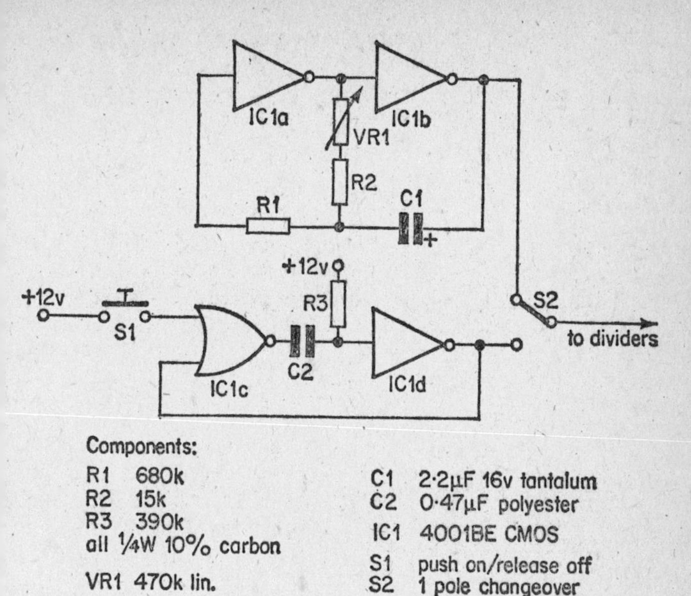

Fig. 4.2 Circuit diagram of clock + one step

and the trigger pulse monostable, and are shown in Figure 4.3

The divider circuit consists of a decade counter (IC2) and 3 NOR gates and an inverter (IC3). The counter counts up to the number set by the rotary switch S4, at which point a pulse is sent to the output while the counter is reset. Thus the output frequency is that of the input divided by the number set by S4. S3 simply serves to by-pass the divider to give the effect of 'divide by one'.

IC4b and IC4c form a monostable to convert the pulse from the divider into a pulse with an 'on' time variable between 50 mS and 3 s. The time constant is set by the two combinations of C3/R4/VR2a and C4/R5/VR2b, which give a well-defined period if VR2 is dual-gang. Since the circuit is activated by the trailing edge of a pulse, IC4a achieves the opposite effect, so that we can trigger an envelope shaper at

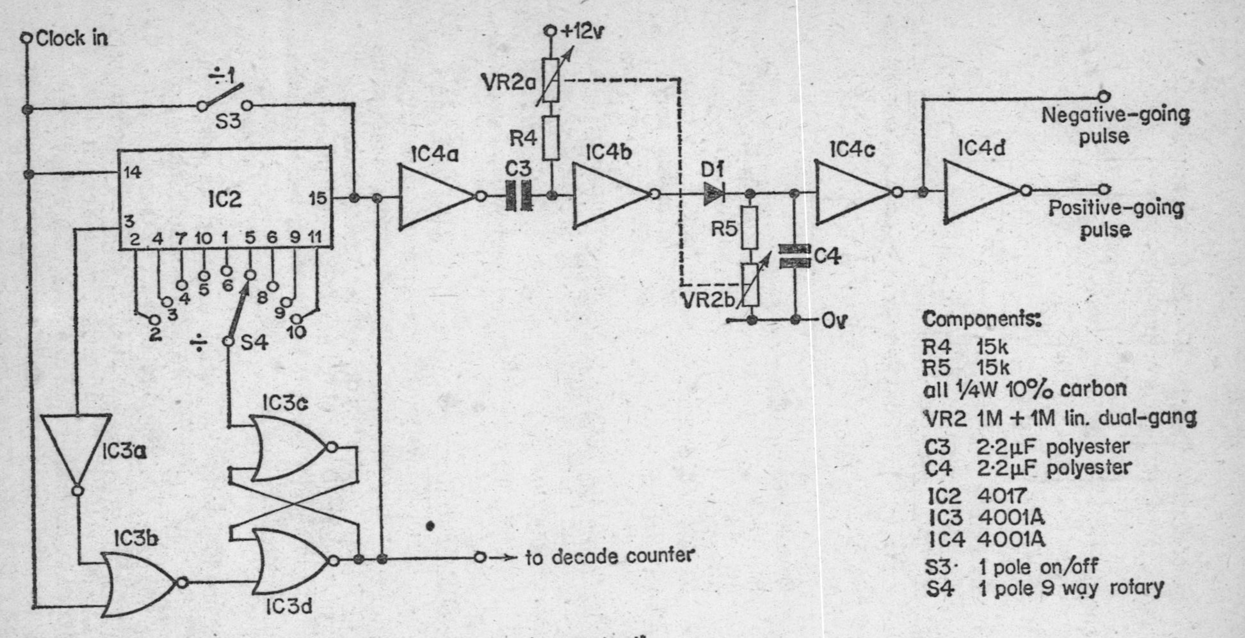

Fig. 4.3 Circuit diagram of divider + monostable (2 required)

the very beginning of a note. IC4d gives us the choice of a positive – or negative-going output pulse.

Figure 4.4 gives the circuit diagram of the decade counter, pots and output amplifier. Note that only one channel is shown, whereas two will be required.

The decade counter (IC5) works by splitting up a clock input so that each of the counter's 10 outputs goes high one after another. The switch from one output to the next occurs on the positive edge (i.e. the leading edge) of the incoming pulse, so we do not have to square up the clock pulse after it leaves the divider.

When an output of the decade counter goes high, one of the pots (VR3–12) becomes active, and the voltage set at the slider is taken via D12–21 to the non-inverting amplifier IC8. The gain of this amplifier is varied by VR13 and has a range of roughly 5:1. VR14 sets the final level sent to the VCO. The reason for having both a gain and level control, is so that the voltage ranges of both the sequencer and VCO can be matched, and thus full use will be made of both pieces of equipment.

D2–11 are l.e.d. indicators to show which particular pot is on at any one time. They are driven by inverters in IC6 and IC7.

S5 connects the reset pin of the counter to any of its outputs so that the sequence length can be adjusted to anything between 1 and 10 notes.

Finally we come to the power supply. The CMOS circuits require minimal current at 12 V, while the 741 amplifiers need a bit more at ±12 V. Although a regulated supply is not absolutely essential, the cost of using I.C. regulators is not a great deal more than using a simpler circuit with discrete components. The recommended circuit is shown in Figure 4.5.

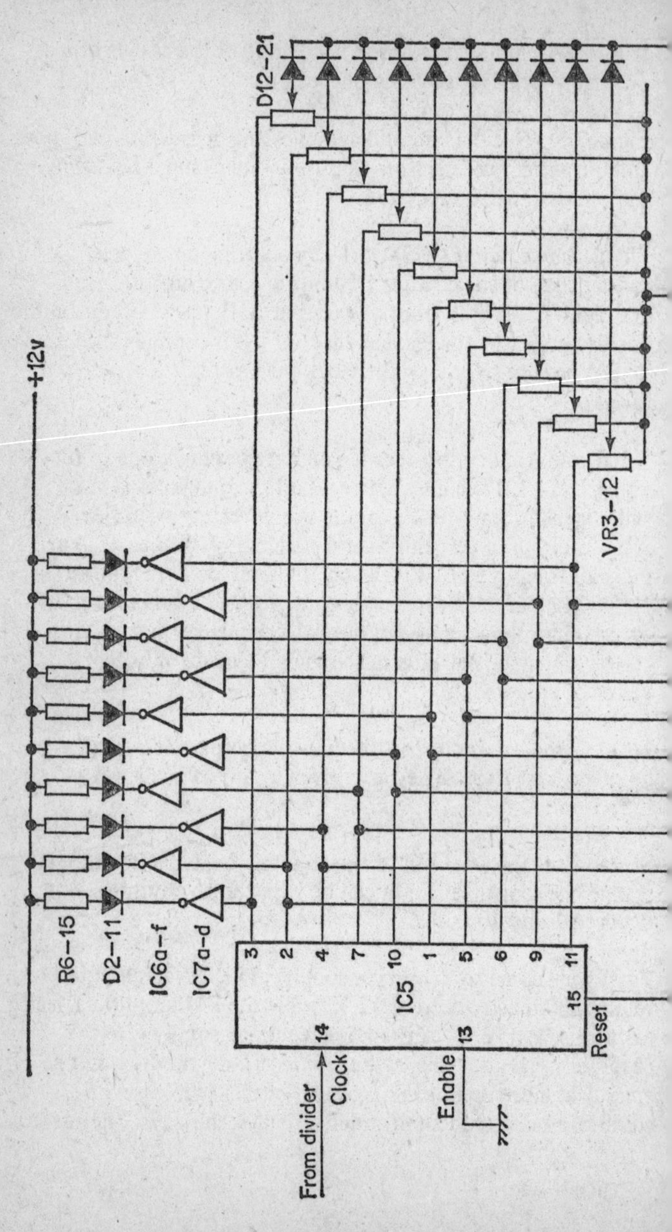
+12v
R6–15
D2–11
IC6a–f
IC7a–d
D12–21
VR3–12
IC5
3
2
4
7
10
1
5
6
9
11
14
13
15
Clock
From divider
Enable
Reset

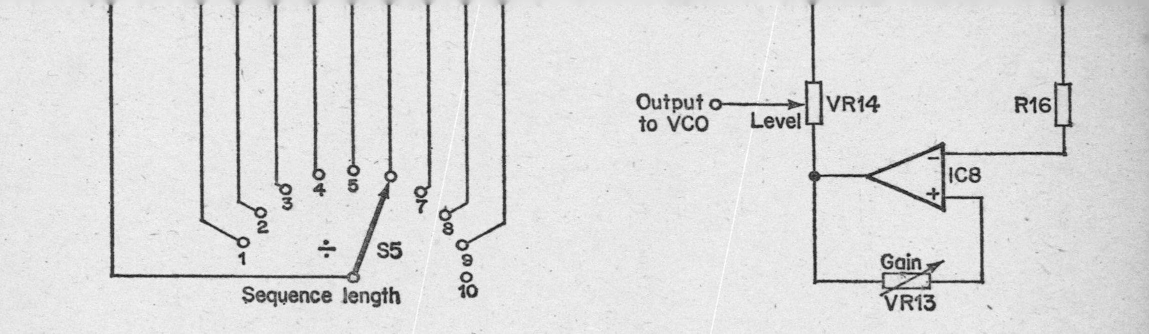

Components:

R6–15	560R
R16	2·2k
all 1/4W 10% carbon	
VR3–12	100k lin.
VR13	22k lin.
VR14	4·7k lin.

IC5	4017
IC6	4049A
IC7	4049A
IC8	741
D2–11	Red L.E.D.s
D12–21	1N914
S5	1 pole 10 way rotary

Fig. 4.4 Circuit diagram of decade counter + pots + output

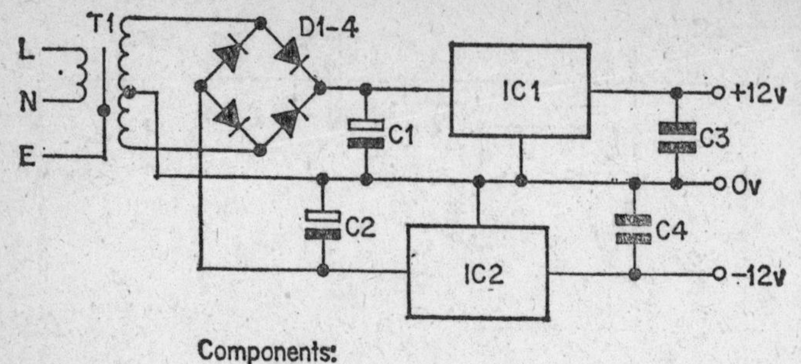

Components:

C1	1000µF 25v elect.
C2	1000µF 25v elect.
C3	0·1µF polyester
C4	0·1µF polyester
IC1	MA78M12UC +12v regulator
IC2	MA79M12UC –12v regulator
D1–4	W01 bridge rectifier
T1	Mains transformer 12–0–12v at 0·5A

Fig. 4.5 Circuit diagram of P.S.U. for sequencer

Construction

3 printed circuit boards are used – 1 for the clock/one-step and p.s.u., and the other two for two channels of divider/monostable/decade counter/output. The copper patterns and board layouts are given in Figures 4.6 and 4.9 inclusive.

The best method of construction for this type of project – or, for that matter, for most types of projects – is to hang everything onto the front panel including, in this case, the power supply. The front panel will have to be suitably large anyway to accommodate all of the pots, and the amount of board-to-controls wiring is going to be astronomical, so it only makes sense to have the boards fixed to the panel.

The construction details here assume that the pots are rotary. Sliders would be very nice, but unless you can get a job lot of them from somewhere, it could turn out to be a bit expensive.

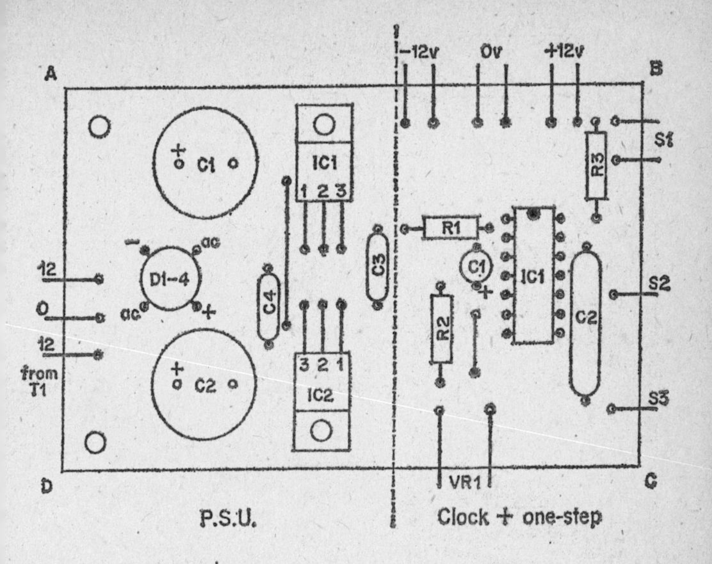

Fig. 4.6 Printed circuit layout for P.S.U. and clock/one-step circuits

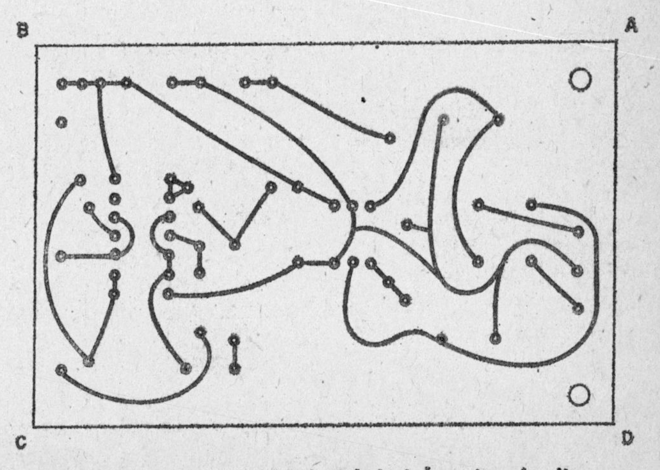

Fig. 4.7 Copper pattern for P.S.U. and clock/one-step circuits

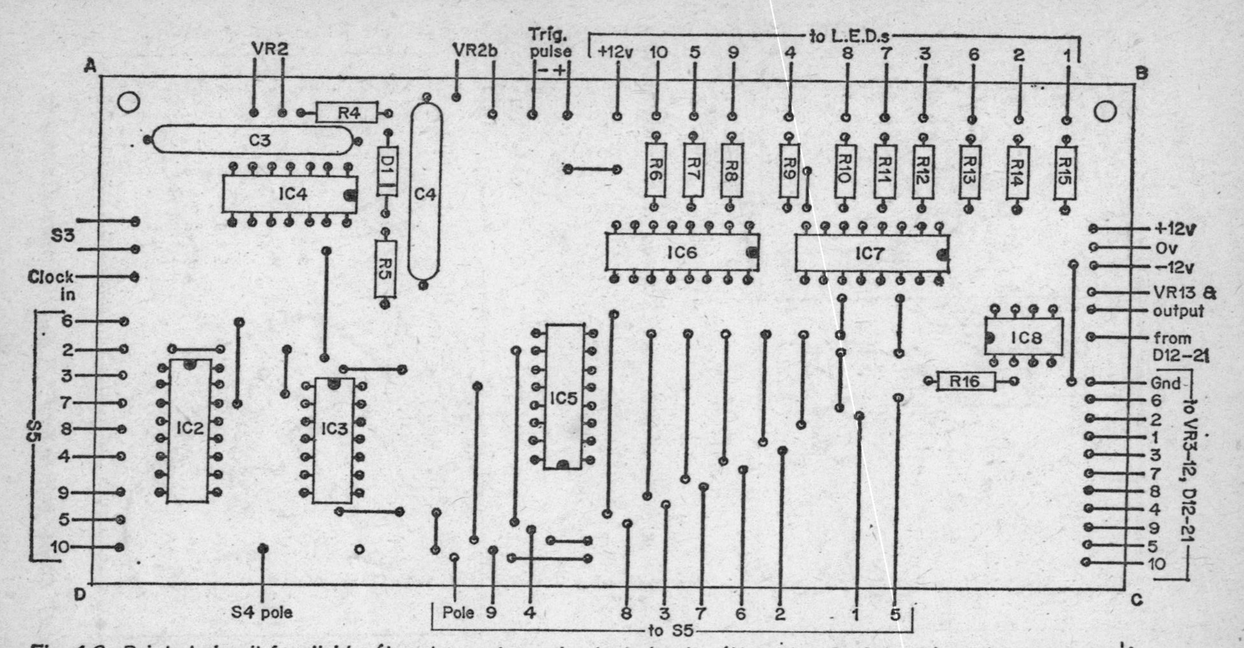

Fig. 4.8 Printed circuit for divider/decade counter and output circuits. (Note that 2 of these boards are required)

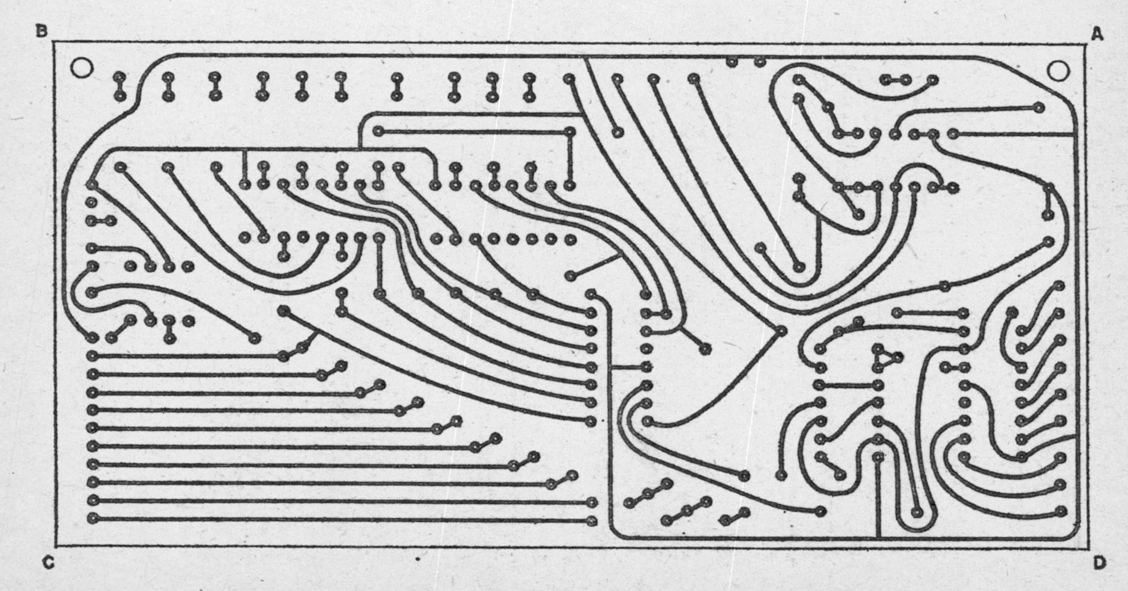

Fig. 4.9 Printed circuit copper pattern for divider/decade counter and output circuits

Figure 4.10 shows how the front of a front panel might look like, while Figure 4.11 takes a look round the back. Dimensions are only approximate, since they depend on what sort of components are available. It is probably best to collect all of your pots and switches and sit down with a large sheet of paper to try and work out what would be a neat arrangement. Then transfer your ideas onto a sheet of 18 s.w.g. aluminium and start drilling.

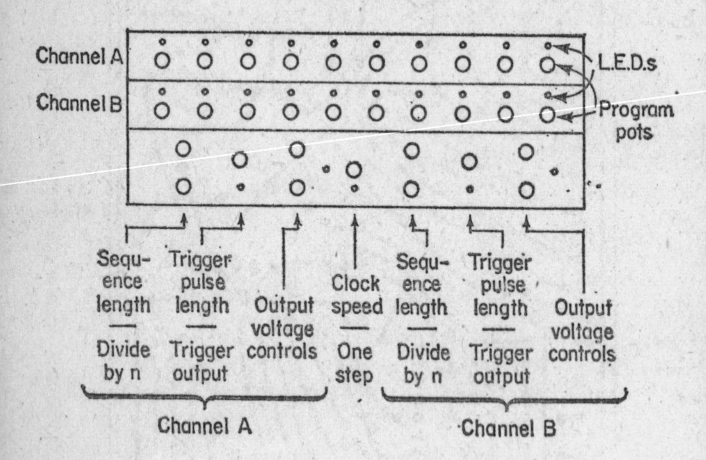

Fig. 4.10 Front panel layout

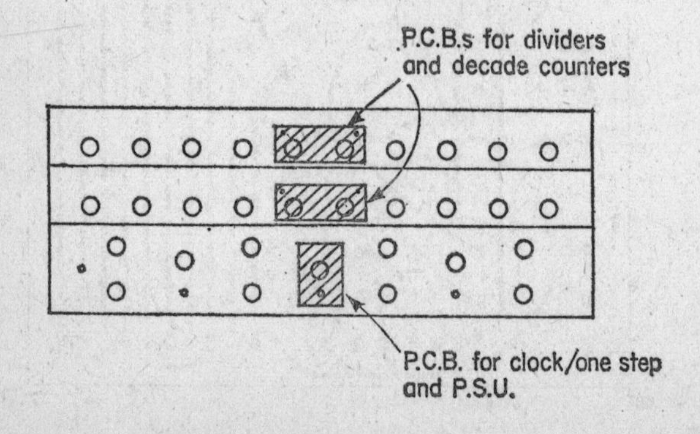

Fig. 4.11 Layout behind front panel

Once you have drilled, painted, lettered the front panel, start making up the boards, beginning with the p.s.u. and clock board. It is a good idea to check this board is working before making up the other two. With all three boards completed and working, the pots and switches should be attached to the front panel. Now the wiring can be carried out, following the circuit diagrams. If suitably coloured wires are used then this part of the proceedings should be straightforward. The boards can now be fixed to the panel and the sequencer checked out.

Using the Sequencer

The basic method of operation should, by now, be fairly apparent. You can soon get used to it by connecting up a couple of VCO's and envelope shapers, as in Figure 4.12, so that each channel has an independent output. By experimenting with clock speed, division number, sequence length and the pots, you will soon be able to see what the sequencer is capable of. To really illustrate how the sequencer works, though, we will see how we can generate background rhythm and bass accompaniment to a piece of music.

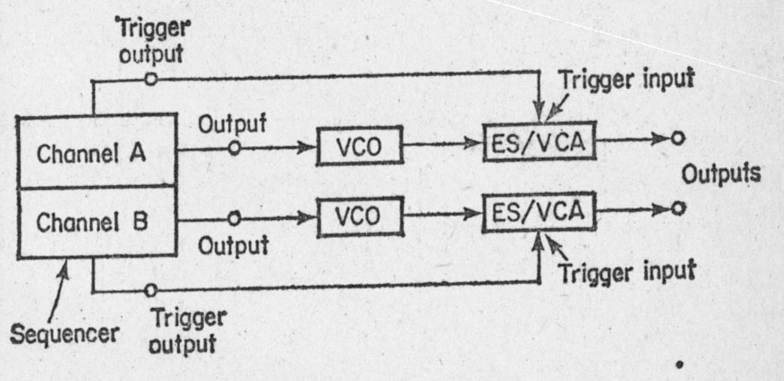

Fig. 4.12 Arrangement of sequencer with VCOs and envelope shapes

Figure 4.13 shows a chart with two lines of music, two lines underneath each one for each channel of the sequencer. The tune is the first part of a traditional Irish jig called 'Dingle Regatta'. It has a good rhythm to it and a simple structure, so it will be easy for us to fit some sequencer patterns into it. There is no need to worry if you know nothing about musical theory – a short explanation follows.

The notes are divided into bars of equal length (timewise). There are three types of notes used and these are also shown in

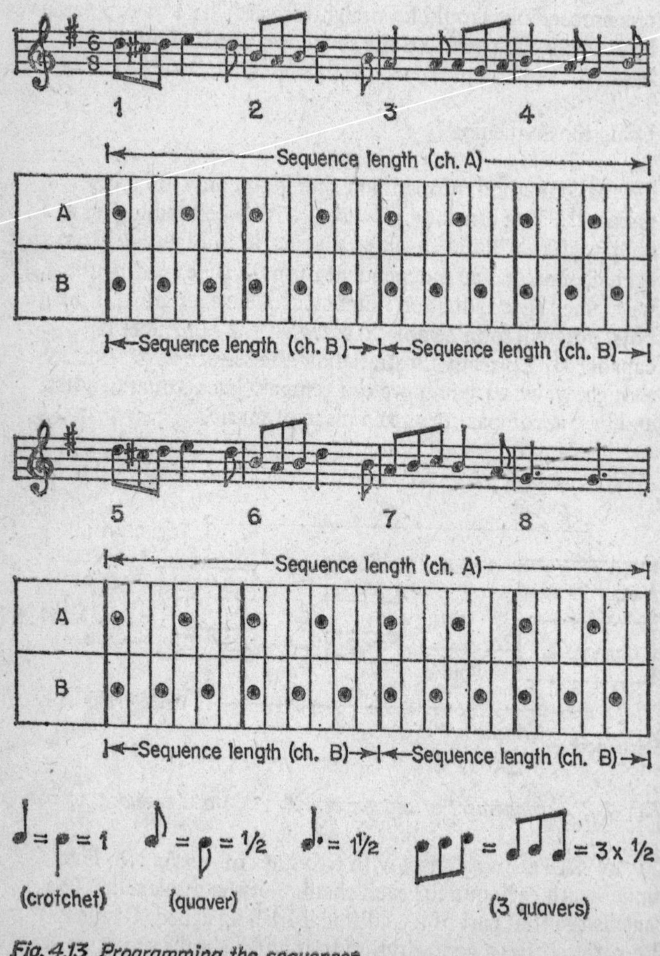

Fig. 4.13 Programming the sequencer

the diagram, together with a time value. The time signature 6/8 at the start of bar 1 means that there are 6 quavers in every bar, and two beats to a bar. This means that there is a beat on the first and fourth quavers in every bar, remembering to treat a crotchet as two quavers.

To program the sequencer, the first thing to do is to set the dividers on each channel. If channel A is going to provide the bass accompaniment and channel B the background rhythm, the pattern of notes or beats will be as shown in the diagram. The bass accompaniment occurs on every beat, and the rhythm runs at 3 notes per bar, equally spaced. The division number for each channel is determined by looking at the frequencies of notes for one channel compared to the other. In this case we see that for two notes in channel A there are three in channel B. Thus, for channel A we need to divide the clock frequency by 3, and for channel B, divide it by 2.

Having set the dividers, we must now set the sequence length. Here we consider for how many notes (up to a maximum 10) we want a sequence to run before starting it again. The best method in this case is to have lengths of 8 notes for channel A and 6 notes for channel B.

The timing is now set and the rest of the setting up is easily done while the tune is played. The first control to adjust is the clock speed. The program pots can then be adjusted so that the pitch of each note begins to sound about right. This is something that cannot be laid down in hard and fast rules, it being dependant on your own preferences.

Of course, these notes are only really for guidance. Once you have mastered the timing controls of the sequencer, you will soon find out how versatile this instrument is. If you find that the sequence lengths or number of channels are becoming a restriction to your creative ideas, then it is possible to add more decade counters and program pots to extend the sequencer.

CHAPTER 5

TWO VCO's.

The VCO is the central element in a synthesiser. It generates the sound which is subsequently processed by circuits such as filters and envelope shapers. The principle of operation of a voltage-controlled oscillator is that the frequency of the output is determined by a voltage level at the input. The relationship between voltage and frequency is normally either linear or logarithmic. The advantage of using a logarithmic VCO is that, when running the VCO from a keyboard which outputs linear voltage increments, the frequencies will match closely with an equal temperament scale.

Two VCO's are described here – one uses a linear voltage/ frequency relationship, the other a logarithmic one. Both would be suitable as sound generators within a synthesiser, and would also work well with other circuits described in this book, notably the sequencer described in the previous chapter.

VCO 1

This first VCO makes use of a readily available function generator I.C., the 8038. This has a frequency range from 0.001 Hz to 1 MHz and has three waveform outputs – sine, triangle, and square wave. In addition, the harmonic content of the sine wave can be adjusted. The circuit diagram for this VCO is given in Figure 5.1. Manual frequency control is via VR1, while VR2 adjusts the sine waveform – this is useful for providing a variety of different tones. All three waveforms are mixed using VR3–5 before being summed and amplified by IC3. A wide variety of waveforms are therefore available from this circuit. Without IC2, a negative-going control voltage will cause an increase in output frequency. If you require a positive-going control voltage, IC2 is a unity gain voltage inverter that will do this.

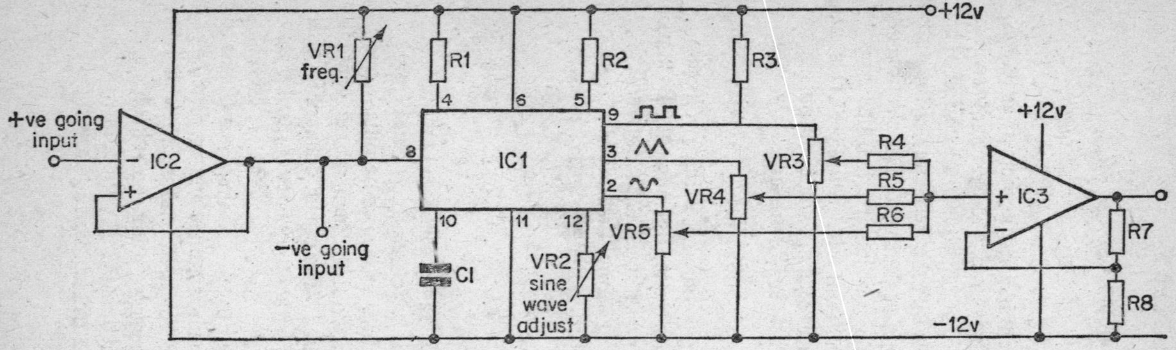

Components:

R1	100k
R2	100k
R3	1M
R4	22k
R5	22k
R6	22k
R7	22k
R8	10k

all 1/4W 10% carbon

VR1	100k lin.
VR2	100k lin.
VR3	10k lin.
VR4	10k lin.
VR5	10k lin.
C1	3300pF polystyrene
IC1	8038 function generator
IC2	741
IC3	741

Fig. 5.1 Circuit diagram of VCO1

Construction

This is straightforward and should not cause any problems. A printed circuit layout and pattern are shown in Figure 5.2. The circuit can be housed in a case by itself or in the same case as other synthesiser modules. Chapter 7 gives details of a possible case construction and also the p.s.u. for this circuit.

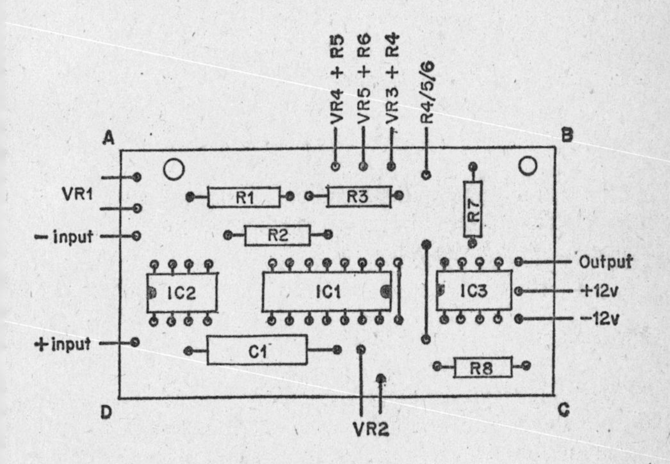

Fig. 5.2a Printed circuit layout for VCO 1

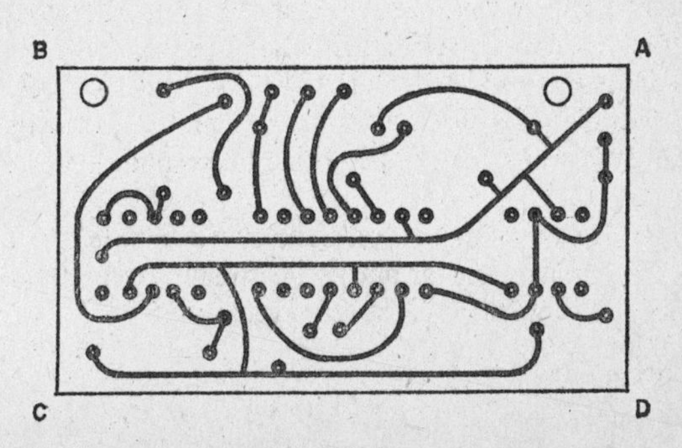

Fig. 5.2b Copper pattern for VCO 1

5

VCO 2

This VCO has a logarithmic-voltage/frequency relationship generated by Tr1 in Figure 5.3. A control voltage varying from 0 to +5 V will cause a current to flow through Tr1. This current will flow through either IC4a or IC4b, depending on the output state of the CMOS dual comparator IC2. Let us assume that it is high, and thus IC4a is on and IC4b is off. IC1, acting as an integrator, ramps in the positive direction. The comparators in IC2 switch at 1/3 Vcc and 2/3 Vcc respectively, which is 4 V and 8 V, IC2 goes low, which turns off IC4a, and turns on IC4b. IC1 then begins ramping in the negative direction until it reaches 4 V, at which point the output of IC2 goes high again, and the cycle repeats.

The slope of IC1's ramp depends upon C1 and the current drawn by Tr1, and so it is easy to see that the output frequency is dependent upon the voltage at R1.

Construction

All the components are assembled on one printed circuit board, which is shown in Figure 5.4. Construction is straightforward and should cause few problems, but be sure to take the usual precautions where the CMOS I.C.'s are concerned – it is advisable to use sockets for the more expensive ones.

The circuit requires no setting-up and can be put into use immediately. Separate sockets should be provided for the triangle and square wave outputs. The control input can be fed by a number of different sources – a simple pot, a keyboard, or even another VCO to obtain cross modulation.

The use of these two circuits on their own is simple and requires no explanations or hints. The final chapter deals with VCO's as part of a large synthesiser system.

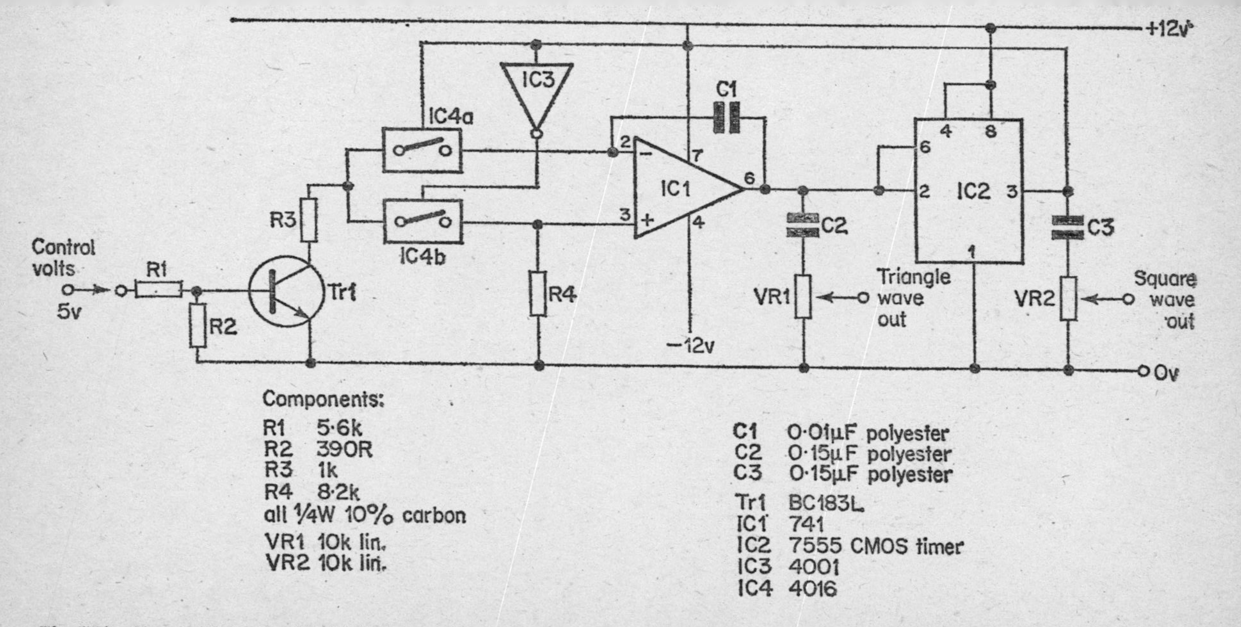

Components:

R1	5·6k
R2	390R
R3	1k
R4	8·2k

all ¼W 10% carbon

VR1	10k lin.
VR2	10k lin.

C1	0·01µF polyester
C2	0·15µF polyester
C3	0·15µF polyester

Tr1	BC183L
IC1	741
IC2	7555 CMOS timer
IC3	4001
IC4	4016

Fig. 5.3 Circuit diagram of VCO2

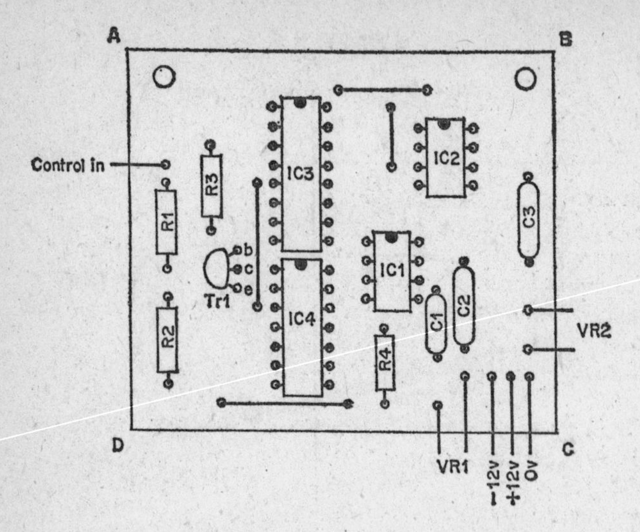

Fig. 5.4a Printed circuit layout for VCO 2

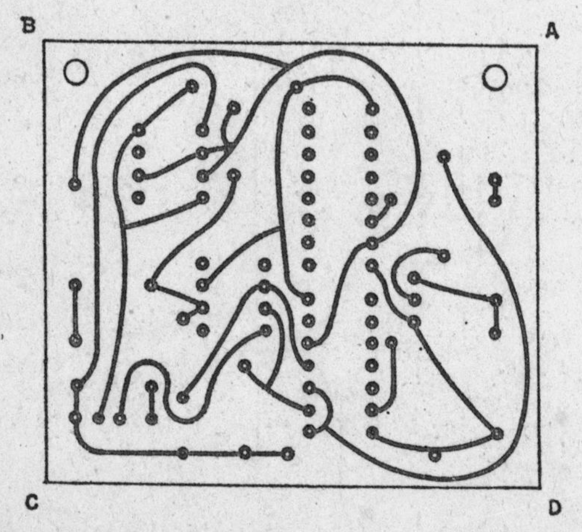

Fig. 5.4b Copper pattern for VCO 2

CHAPTER 6

A.D.S.R. ENVELOPE SHAPER

An envelope shaper is basically an amplitude modulator. In a synthesiser it is usually the last stage before final mixers and amplifiers and its job is to impart an envelope to the sound. Without an envelope shaper the output would be a continuous tone. Figure 6.1 shows just what the shaper does to the sound.

There are two types of envelope shaper, namely A.D. and A.D.S.R. The top diagram in Figure 6.1 shows the operation of an A.D., or attack-decay, shaper. On receiving the necessary trigger pulse, the envelope embarks on the attack section. When it reaches a peak, it decays away to zero. Both the attack and decay rates are variable.

In the A.D.S.R. shaper, shown in the diagram below, a more complex waveform is generated. The trigger pulse is often generated from a keyboard, and when the envelope shaper receives the leading edge of the pulse, it begins on its attack slope at a preset rate. On reaching a peak, the level decays at a rate that can also be adjusted, but instead of falling to zero it levels off at an adjustable sustain level. The envelope stays at this level until the trigger pulse ends, when the level dies away to zero at a rate set by the 'release' control. When the

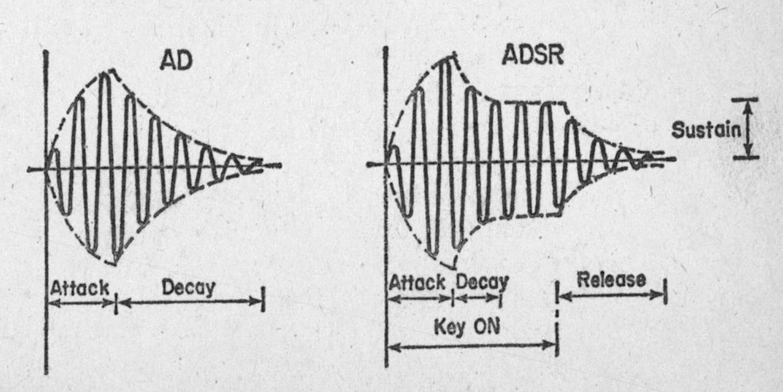

Fig. 6.1 Diagrams of AD and ADSR envelopes

output from a VCO is amplitude-modulated with this envelope, very good imitations of most string instruments can be achieved.

The envelope shaper described here is of the A.D.S.R. type just described. It comes complete with an amplitude modulator and can be used in conjunction with either of the VCO's described in the last chapter, or indeed most other VCO's.

Circuit Operation

A complete circuit diagram is given in Figure 6.2. Referring to the timing diagram in Figure 6.3 will assist in understanding the sequence of events.

Before the start of an envelope S1 is off, IC1a and b are on, and IC1c is off. IC1 is a quad bilateral CMOS switch, which is a set of four separate switches. If a logic 1 is applied to the control input of a switch, it will turn on. When S1 closes, or the leading edge of a trigger pulse reaches the input, IC1a turns off and C1 begins charging via VR1, the Attack control. As the level at C1 reaches +7 V the inverter IC2c changes state and the latch (IC2a and IC2b) changes state. This turns IC1b off and IC1c on, which allows C1 to discharge through the Decay control VR2. However, it will only discharge until it reaches the voltage level set by VR4, the sustain control, and there it will stay for as long as S1 is held closed (or the input trigger pulse is held high). When eventually the input goes low, IC1a switches on again and C1 discharges completely through VR3, the sustain control.

The generated DC envelope is taken to a buffer with a gain of 3, built around op-amp IC3. VR5 adjusts the output level and hence the depth of modulation of the voltage-controlled amplifier which follows. This is based on an inexpensive chip, the MC3340P. It comes in an 8-pin d.i.l. package and can attenuate an applied signal by up to 90 dB. VR6 controls the final signal output level.

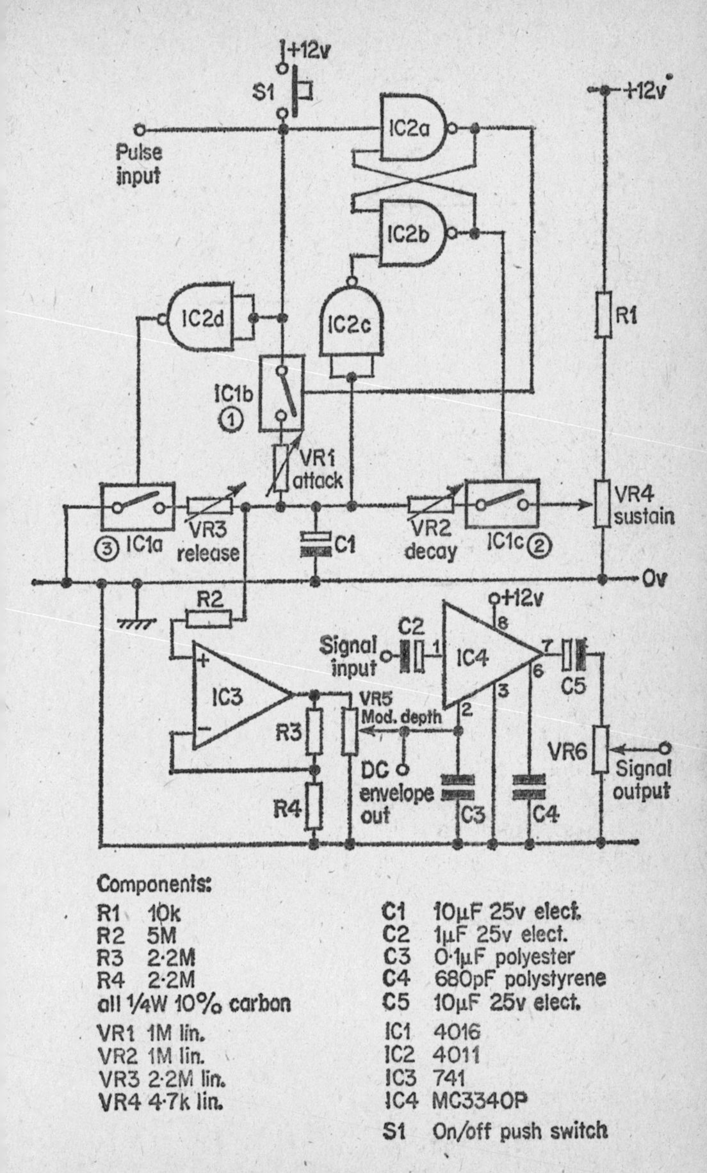

Fig. 6.2 Circuit diagram of ADSR envelope shaper

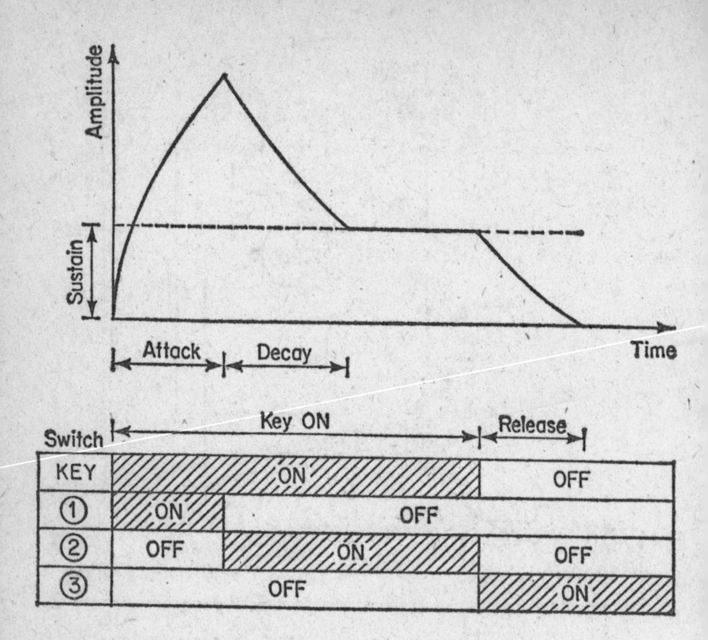

Fig. 6.3 ADSR timing diagram

Construction

A printed circuit layout and copper pattern for the envelope shaper is given in Figure 6.4. Construction and setting-up are straightforward and should cause few problems. Take care, when soldering I.C.'s in, not to bridge copper tracks with stray pieces of solder. It is a good idea to use sockets for the I.C.'s.

Details of housing the circuit are not given here, since it is expected that it will probably form part of a synthesiser or other sound effects unit and will therefore fit into a larger case. Of course, the circuit can be housed by itself, but some thought must be given to the power supply. The p.s.u. is described in the next chapter and is intended to be used to run a collection of synthesiser circuits, including this envelope shaper.

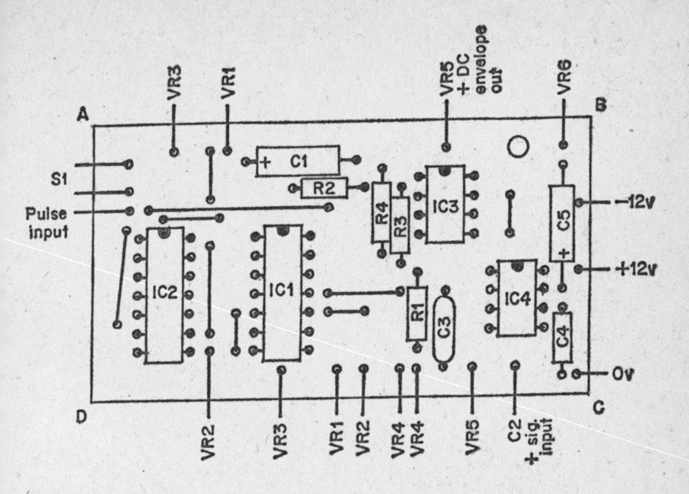

Fig. 6.4a Printed circuit layout for envelope shaper

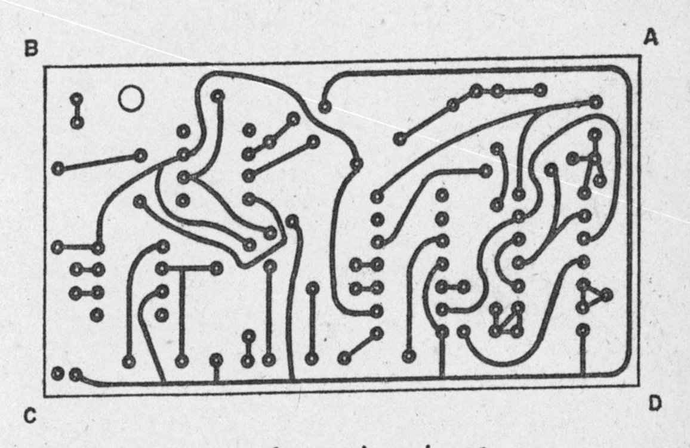

Fig. 6.4b Copper pattern for envelope shaper

Using the Envelope Shaper

Using this circuit is a lot simpler than the impression you may get by looking at an array of six knobs. The circuit is triggered by one of two means:

a) Manual push switch S1

or

b) Trigger pulse: logic 0 = 0 V
logic 1 = +12 V

The output of the circuit can be connected to the input of an amplifier, while a VCO can provide the signal input. Then, by experimenting with all of the various controls, you will find out how the different envelopes sound. For instance, setting a fast attack, a fairly short decay, mid-value sustain, and a long release time will give a good imitation of a piano, providing of course, that the VCO is generating the right waveform.

A violin imitation, on the other hand, is achieved with a long attack and decay, low sustain and short release. Naturally, the envelope shaper is not intended just for imitating – an infinite variety of new envelopes can be generated.

CHAPTER 7

POWER SUPPLY UNIT

This power supply is intended to run a collection of synthesiser circuits such as the VCO's and envelope shaper described earlier. All the circuits require a standard ±12 V supply, preferably regulated. Since regulator chips are now available for a variety of voltage and current requirements, they have been used for this particular p.s.u.

The circuit diagram is shown in Figure 7.1. Operation is straightforward, since the regulator chips take care of most of the processing.

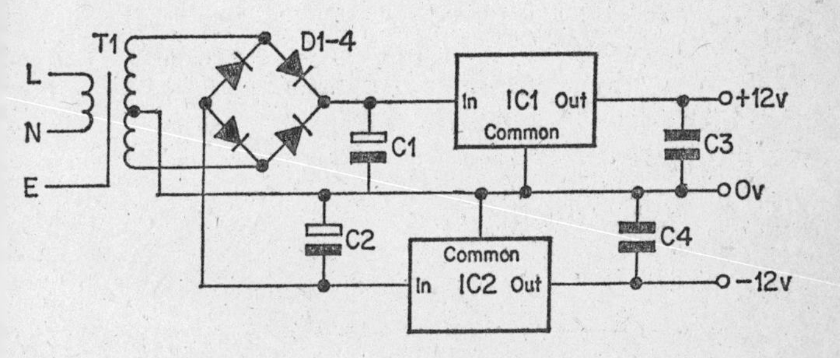

Components:

C1 1000µF 25v elect.
C2 1000µF 25v elect.
C3 0·1µF polyester
C4 0·1µF polyester
IC1 µA7812UC +12v regulator
IC2 µA7912UC –12v regulator
D1–4 W01 bridge rectifier
T1 Mains transformer 12–0–12v at 1A

Fig. 7.1 Circuit diagram of synthesiser power supply

Construction

A printed circuit board layout and copper pattern is shown in Figure 7.2, which carries all of the components except the regulator chips and the transformer. These are mounted, together with the circuit board, on a piece of aluminium, so that the metal acts as a heatsink for the regulators and also as a screen for the transformer. Figure 7.3 gives an idea of what this arrangement should look like. The complete p.s.u. is thus easily fitted into the corner of a large case containing other synthesiser modules. The best method of construction in this situation would be to fit the synthesiser circuits (VCO's, VCF's, etc.) in the form of modules onto the front of the box so as to form a front panel. Each module would be a complete unit with circuit board, pots, switches and input/output sockets mounted on a panel. The power connection would be in the form of a multiway plug and socket arrangement, whereby several sockets would be mounted along the bottom of the box and connect to the p.s.u. In this way, each module can be very easily removed for servicing and repair by unscrewing the panel and pulling out the power connector from the base of the box.

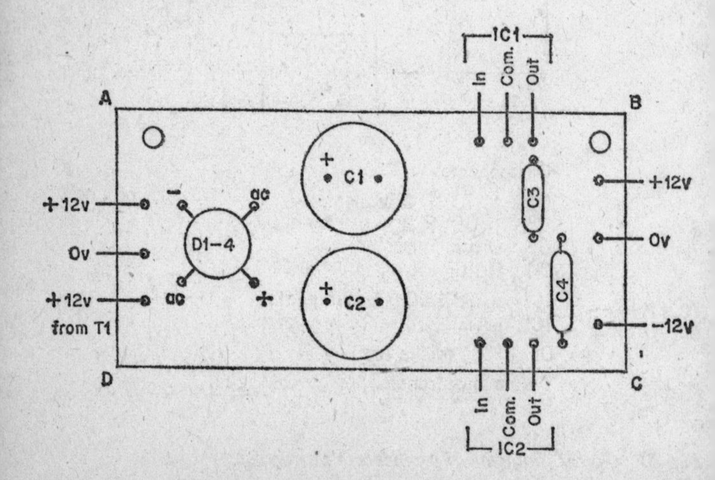

Fig. 7.2a Printed circuit layout for power supply

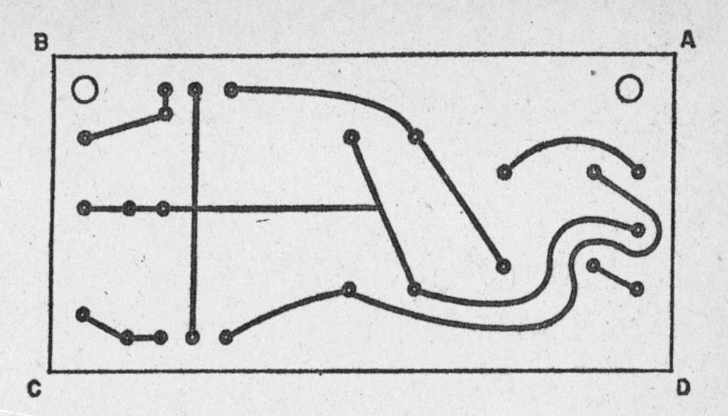

Fig. 7.2b Copper pattern for power supply

Figure 7.3 shows a general arrangement for this useful method of construction.

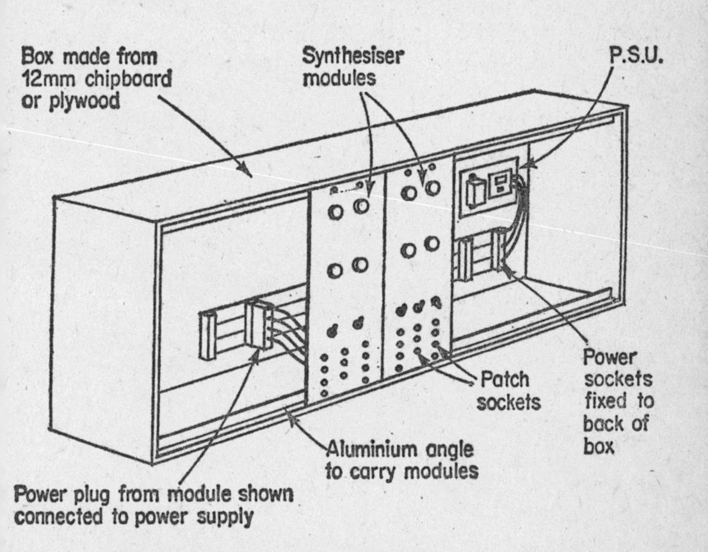

Fig. 7.3 Sketch showing arrangement of synthesiser modules in box

CHAPTER 8

PUTTING IT ALL TOGETHER

The object of this final chapter is to show how the various circuits described in this book can be made to work together to produce a versatile piece of sound-generating equipment. It goes on to show briefly how a synthesiser or sound generators would fit into an electronic music studio.

Synthesisers

A synthesiser is a set of sound-generating and sound-treatment circuits housed in one box. The individual circuits can be connected together in a variety of ways (otherwise known as patching). The difference between large and small synthesisers is merely the number of times that the basic circuits are duplicated. A small synthesiser would have a minimum of two VCO's, whereas a larger one may have as many as nine or ten.

The VCO can be seen as the starting point in any synthesiser – generally 2 are considered a minimum. Then you need one or two envelope shapers, and you have a very basic synthesiser. The VCO and envelope shaper circuits described earlier can be used together to form such a synthesiser. Of course, you will need the p.s.u. described in Chapter 7 as well.

With just these circuits, a surprisingly large range of sounds and noises can be generated. The normal patching arrangement for 2 of each of these circuits would be as shown in Figure 8.1.

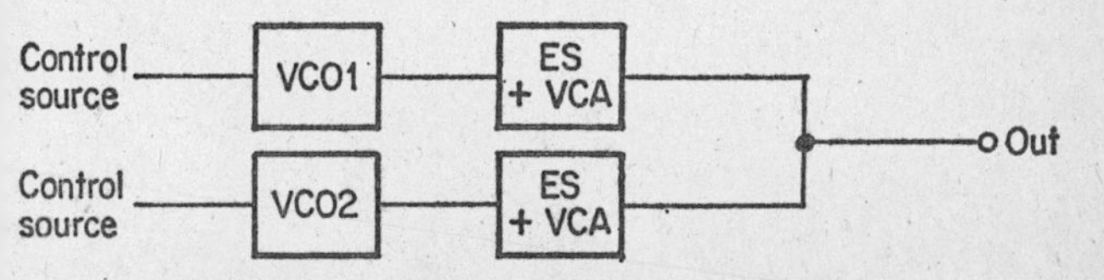

Fig. 8.1 Basic synthesiser set-up

Each pair of VCO and envelope shaper generates its own particular sound, and then the two are mixed together. Since each shaper has its own level control, we can mix the two in varying proportions.

Alternatively, one envelope shaper could be used to accept the output of both VCO's. This results in a single rather than a double sound, but with a much richer tone structure. For instance, with 4 VCO's a rich 4-note chord can be obtained. By selecting the appropriate waveforms from the VCO's and by setting up a suitable envelope, many musical instruments ca be imitated or some new ones 'invented'.

Another useful way of using 2 or more VCO's is to allow them to modulate one another. This is easily achieved by connectin the output of the modulating VCO to the control input of the modulated VCO. Figure 8.2 shows an arrangement whereby th output from two VCO's are used to modulate a third. This arrangement can be used to generate complex vibrato sounds.

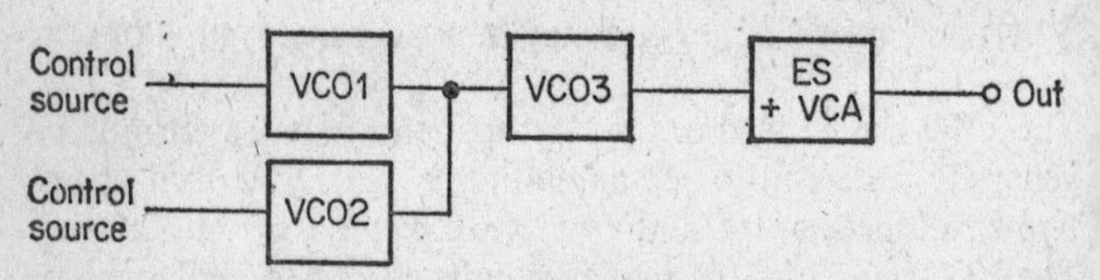

Fig. 8.2 Modulating VCOs

An envelope shaper can also be used to control a VCO. By connecting up a shaper as shown in Figure 8.3, an interesting sound can be heard when manually operating the shaper with the push button. The usual sound envelope will be heard, but at the same time the frequency will rise and fall, reaching its peak at the same point that the envelope reaches its peak.

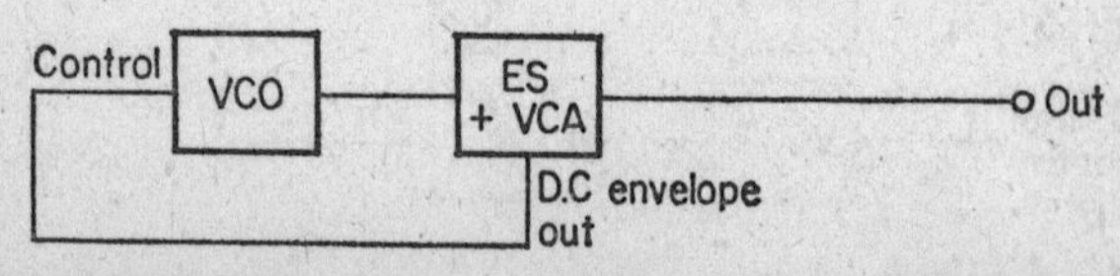

Fig. 8.3 ES controlling VCO

Additional Circuits

Naturally, VCO's and envelope shapers are not the only circuits that you are likely to find in a synthesiser. There are numerous other circuits, which treat the sound in various ways. Probably one of the most common is the voltage-controlled filter (VCF) which is most useful for changing the harmonic content of a sound. In a VCF, which is usually a low-pass type, the controlling input voltage is directly related to the cut-off frequency of the filter. Figure 8.4 shows a common use of a VCF in conjunction with an envelope shaper and a VCO. The control voltage used for the VCO is also used for the VCF so that the pass band of the filter 'tracks' the VCO frequency. This means that the harmonic content stays the same at all frequencies.

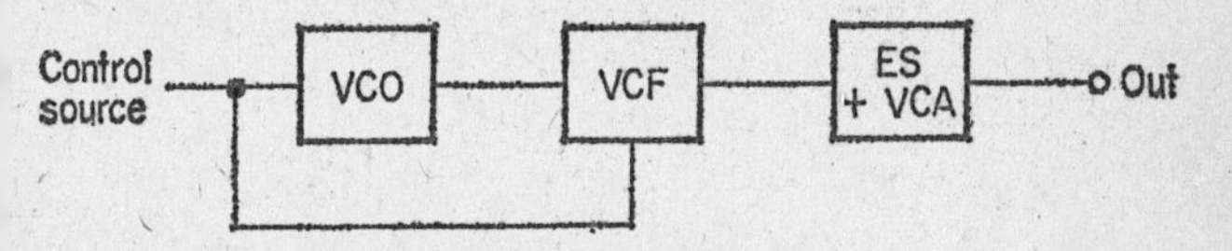

Fig. 8.4 Common VCF circuit

Another frequently used circuit is a noise generator. This produces white noise which contains all frequencies of the audio spectrum and sounds like the noise heard when tuning between radio stations on an FM band. The noise can be used instead of, or mixed with, VCO's and can be treated by the same treatment circuits. Pink noise is white noise whose bass frequencies have been boosted. It can be used in the same way as white noise would be used.

Control Sources

You will have noticed at the front end of the diagrams something called 'Control source'. This is where the basic control voltage for the VCO comes from. In its simplest form this can be a potentiometer giving between 0 and 6 V at its wiper. A better and more usual source is a keyboard. A simple

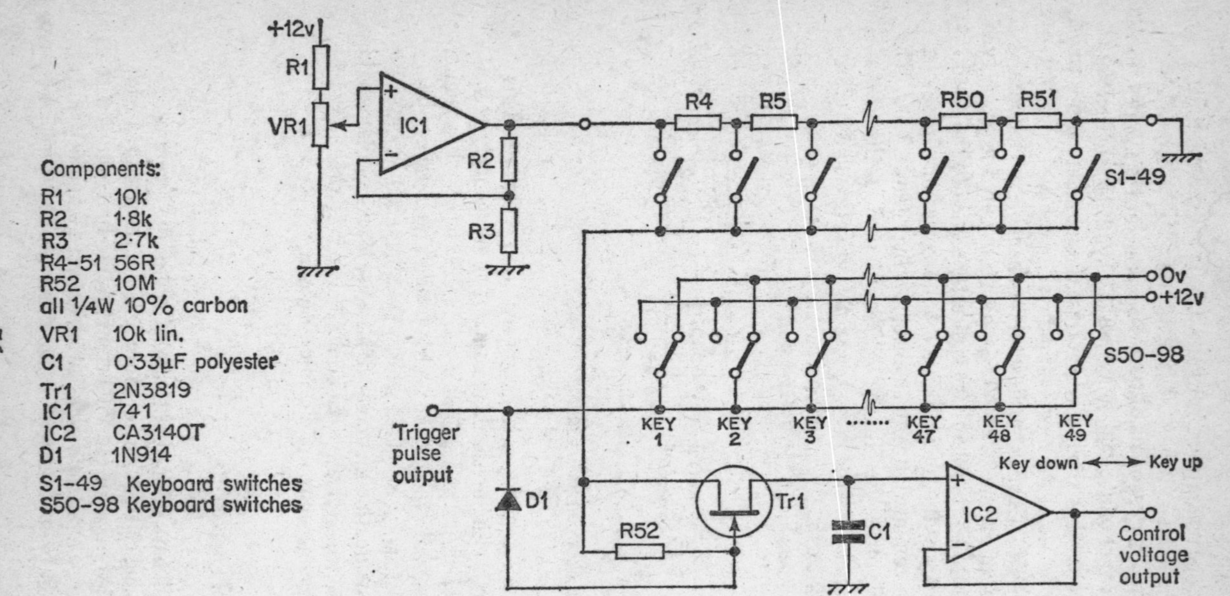

Fig. 8.5 Circuit diagram of keyboard and sample-and-hold circuit

design for one is given in Figure 8.5. It has provision for adjusting the control voltage range using VR1. The keyboard needs to have two sets of contacts — one for the control voltage from the resistor chain, and the other to provide a trigger pulse for the envelope shapers.

The circuit also includes a sample and hold section, which is necessary to remember the last key pressed, and to retain the voltage while the envelope decays away. A printed circuit board for this circuit is shown in Figure 8.6.

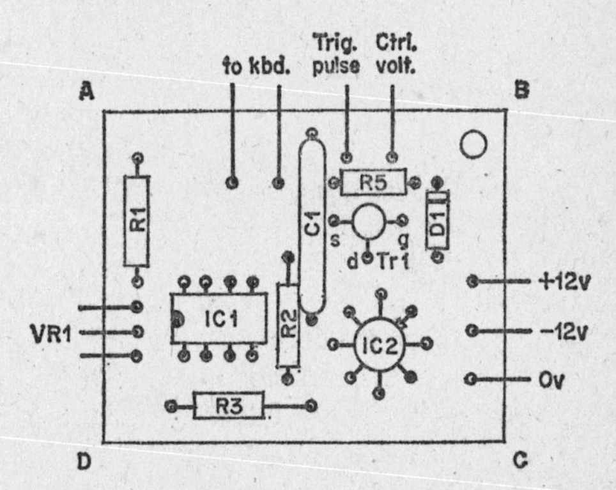

Fig. 8.6a Printed circuit layout for keyboard, sample and hold

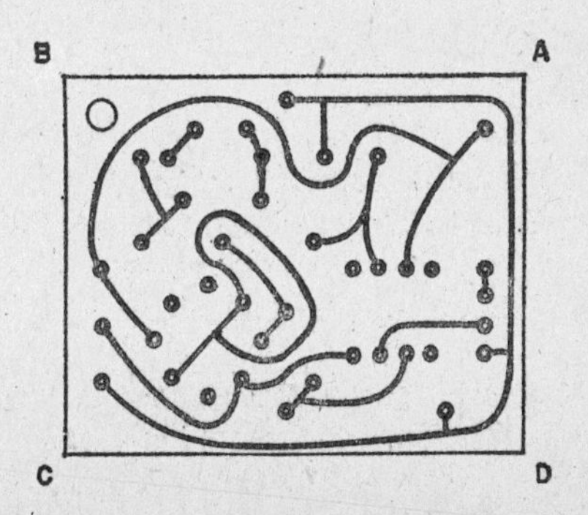

Fig. 8.6b Copper pattern for keyboard, sample and hold

Using a Sequencer with a Synthesiser

Since details of how to use the programmable sequencer with VCO's and envelope shapers have been considered, we shall not go into them again, but suffice it to say that a sequencer is probably the most useful accompaniment to a synthesiser. If you have enough VCO's and envelope shapers to go round, you can use the sequencer to provide the rhythm and bass accompaniment while playing the synthesiser keyboard to provide the melody. Note too that if the output control voltages of the keyboard are added to those of the sequencer, with the sequencer playing through a VCO and a shaper, you can change the key of each sequence by holding down various keys on the keyboard for the duration of each sequence.

Using a Delay Line with a Synthesiser

Any of the various delay line applications described in Chapter 3 can be used with a synthesiser. As well as taking the output of a synthesiser through the delay line, it is also interesting to insert the delay line at various points along the circuit line. For example, taking the output of a VCO and passing it through the delay line, connected up for phasing, will yield a variety of sounds. Remember that the clock generator on the delay line can be voltage-controlled too. This facility enables circuits like an envelope shaper to control the delay time.

An Electronic Music Studio

Although the synthesiser and its accompanying sound effect units are the central units in an electronic music studio; there are some more, very vital, items that you will need to start composing and producing electronic music.

First we shall see what sort of room we need for our activities. Perhaps at this point some of you are thinking that this is rather hypothetical because you would never be able to have a whole room set aside as a studio. Nevertheless the following hints could be applied to a whole room or just an area of a room.

The sort of area that you will need is about 3 m. x 4 m. Thought should be given to acoustics as well – a carpeted floor cuts down on reverberation, but if this is still a problem, try hanging curtains on the walls. Furniture-wise, you need three reasonably-sized tables, a couple of shelves in appropriate places to hold your amplifier and speakers, and a chair from which you can comfortably reach all of the various knobs and switches on the equipment which is laid out on the tables. A swivel chair is the ideal for this purpose.

Equipment

Apart from the synthesiser itself there are a number of other items that you will need, and these are set out below.

1) **Tape recorder(s).** These are necessary if you want to retain your efforts, or if you want to build up complex compositions by multi-recording techniques. For the latter, the ideal is to have two stereo half-track open reel recorders, although stereo cassette recorders can be used instead. A simple mono cassette recorder has limited use, however, but this does not necessarily mean that you cannot do anything with it.

2) **Amplifier and speakers.** Some of you may be tempted to use the family hi-fi, but beware! Domestic hi-fi equipment is not intended for use with continuous pure tones, such as may emanate from synthesisers. Instead you really need a high power stereo amplifier with appropriate speakers. This need not be as expensive as you may think. All you really need are the actual power amplifier modules – there is no need for an elaborate pre-amp at all. About 30 W per channel or more is a good figure to aim at, although if you use a lower rating, check the rating for continuous sine wave inputs and make sure that your speakers are rated above the amplifier's output by at least 30%.

3) **Mixer.** This is the item that replaces the pre-amp in the amplifier. It takes all of the various inputs, adds them together in the proportions you want, and then adjusts the signal for overall level and overall tone content. For our studio we need a mixer with a stereo output, 6–8 inputs (the more the merrier, but more expensive), and VU meters on each output channel. On each input channel we need a channel fader, pan pot, pre-fade listen facility and overload indicator.

4) **Miscellaneous.** This category covers various small items which you could not go without. Leads, for instance, made up in various lengths and with various plugs never seem useful until one Sunday afternoon when you want to connect this piece of equipment to that piece of equipment and you haven't enough leads to do it.

   A small tool kit might be useful to have around, as well as a supply of fuses for the various equipment. On the subject of mains power supply, most rooms do not have half a dozen sockets so it is wise to use a distribution board with 4 sockets on it. This way, only one plug is used in the wall socket and one 13 A fuse covers all the equipment.

Equipment Layout

Figure 8.7 shows an example of how the studio should be laid out. Most of the equipment is situated on the tables around you, with the mixer in the middle, since this controls the total output to the amp. or tape recorders. The amp. and speakers are not kept on the table – the speakers for obvious reasons, the amp. because all volume and tone adjustments can be made from the mixer.

Be sure to keep wiring tidy and around the back of the equipment. Wires trailing across floors should be taped down or otherwise fixed.

Hopefully, these hints will help you create your own electronic

BP81 ELECTRONIC SYNTHESISER PROJECTS
CORRIGENDA & AMENDMENTS

Page 10, Fig. 2.6: 4.7k resistors should be inserted with ends of VR1 and VR6. Add D1–3 in series with S7–9, cathodes towards these switches. Add to parts list D1–3 1N914.

Page 33, Fig. 3.12: R1 should read 330R not 330k.

Page 59, Fig. 5.2b: Copper pattern should be modified as shown:

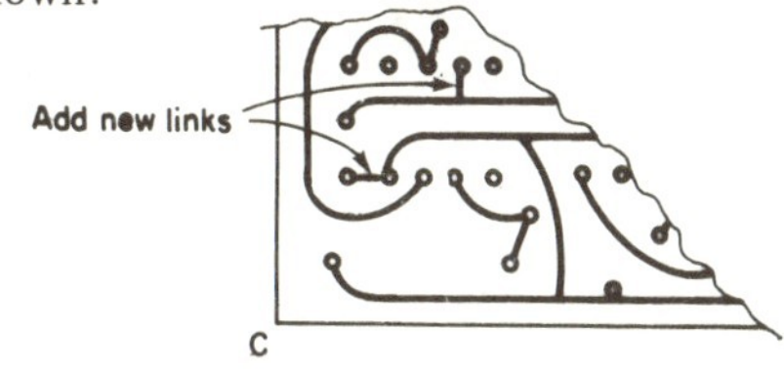

Page 61, Fig. 5.3: IC3 should read 4049 not 4001.

Page 65, Fig. 6.2: Values of VR5 and VR6 have been omitted. VR5 100k lin, VR6 25k lin.

Page 67, Fig. 6.4b: Copper pattern should be modified as shown.

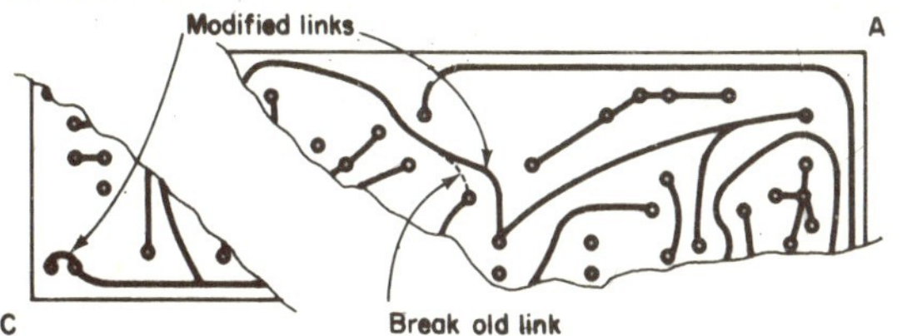

Page 76, Fig. 8.5: Insert one 10k resistor between switches S50–98 and 0v.

Page 77, Fig. 8.6a, 8.6b: Diagram and copper pattern modified as shown:

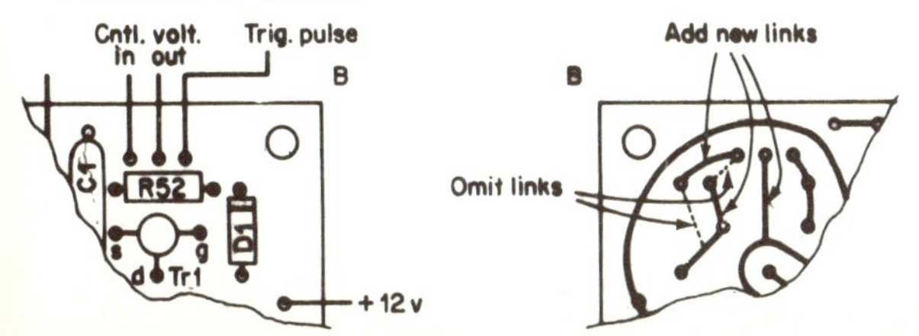

music studio. Once you are set up, the rest is up to you. It is a good idea to experiment fully with your sound-generating and sound effect circuits to find out what you can get them to do. When you find that the speakers are emitting some interesting noises, it's time to get them down on tape and start composing your own electronic music.

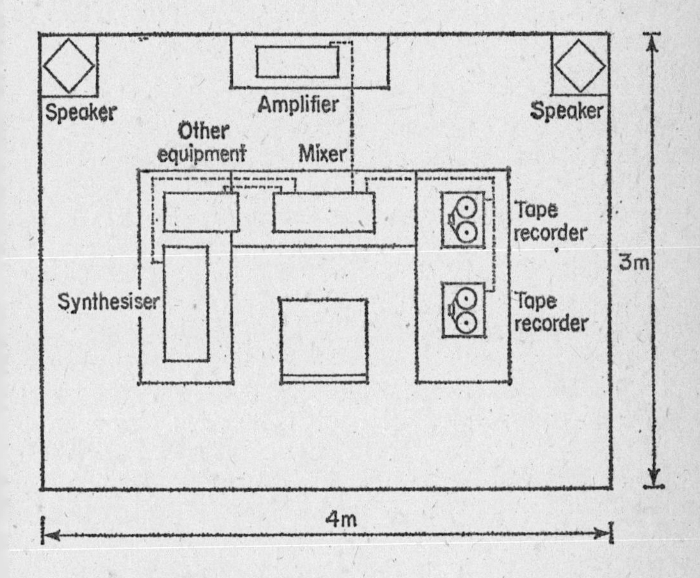

Fig. 8.7 Studio layout

Notes

Notes

Notes

Notes

Notes

OTHER BOOKS OF INTEREST

BP51: ELECTRONIC MUSIC AND CREATIVE TAPE RECORDING
AUTHOR: M. K. BERRY **PRICE: £1.25**
ISBN: 0 900162 72 4 96 Pages

Electronic music is the new music of the Twentieth Century. it plays a large part in "pop" and "rock" music and, in fact, there is scarcely a group without some sort of synthesiser or other effects generator.

It is possible with relatively simple apparatus to create complete compositions using simple electronic and, sometimes, non-electronic musical sources.

This book sets out to show how Electronic music can be made at home with the simplest and most inexpensive of equipment. It then describes how the sounds are generated and how these may be recorded to build up the final composition.

For the constructor, several ideas are given to enable him to build up a small studio including a mixer and various sound effects units.

BP74: ELECTRONIC MUSIC PROJECTS
AUTHOR: R.A. PENFOLD **PRICE: £1.75**
ISBN: 0 900162 94 5 112 Pages.

Although one of the more recent branches of amateur electronics, electronic music has now become extremely popular and there are many projects which fall into this category, ranging in complexity from a simple guitar effects unit to a sophisticated organ or synthesiser.

The purpose of this book is to provide the constructor with a number of practical circuits for the less complex items of electronic music equipment, including such things as Fuzz Box, Waa-Waa Pedal, Sustain Unit, Reverberation and Phaser-Units, Tremelo Generator etc.

The text is divided into four chapters as follows; Chapter 1: Guitar Effects Units; Chapter 2: General Effects Units; Chapter 3: Sound Generator Projects, Chapter 4: Accessories.

205: FIRST BOOK OF HI-FI LOUDSPEAKER ENCLOSURES

AUTHOR: B. B. BABANI **PRICE: 95p**

ISBN: 0 900162 39 2 96 Pages.

This book gives data for building most types of loudspeaker enclosure. Includes corner reflex, bass reflex, exponential horn, folded horn, tuned port, klipschorn labyrinth, tuned column, loaded port and multi speaker panoramic. Many clear diagrams for every construction showing the dimensions necessary.

Please note overleaf is a list of other titles that are available in our range of Radio, Electronics and Computer Books.

These should be available from all good Booksellers, Radio Component Dealers and Mail Order Companies.

However, should you experience difficulty in obtaining any title in your area, then please write directly to the publisher enclosing payment to cover the cost of the book plus adequate postage.

If you would like a complete catalogue of our entire range of Radio Electronics and Computer Books then please send a Stamped Addressed Envelope to:

BERNARD BABANI (publishing) LTD
THE GRAMPIANS
SHEPHERDS BUSH ROAD
LONDON W6 7NF
ENGLAND

160	Coil Design and Construction Manual	1.25p
202	Handbook of Integrated Circuits (IC's) Equivalents & Substitutes	1.45p
205	First Book of Hi-Fi Loudspeaker Enclosures	95p
207	Practical Electronic Science Projects	75p
208	Practical Stereo and Quadrophony Handbook	75p
211	First Book of Diode Characteristics Equivalents and Substitutes	1.25p
213	Electronic Circuits for Model Railways	1.00p
214	Audio Enthusiasts Handbook	85p
218	Build Your Own Electronic Experimenters Laboratory	85p
219	Solid State Novelty Projects	85p
220	Build Your Own Solid State Hi-Fi and Audio Accessories	85p
221	28 Tested Transistor Projects	1.25p
222	Solid State Short Wave Receivers for Beginners	1.25p
223	50 Projects Using IC CA3130	1.25p
224	50 CMOS IC Projects	1.25p
225	A Practical Introduction to Digital IC's	1.25p
226	How to Build Advanced Short Wave Receivers	1.20p
227	Beginners Guide to Building Electronic Projects	1.25p
228	Essential Theory for the Electronics Hobbyist	1.25p
RCC	Resistor Colour Code Disc	20p
BP1	First Book of Transistor Equivalents and Substitutes	60p
BP2	Handbook of Radio, TV & Ind. & Transmitting Tube & Valve Equiv.	60p
BP6	Engineers and Machinists Reference Tables	70p
BP7	Radio and Electronic Colour Codes and Data Chart	35p
BP14	Second Book of Transistor Equivalents and Substitutes	1.10p
BP23	First Book of Practical Electronic Projects	75p
BP24	52 Projects Using IC741	95p
BP27	Giant Chart of Radio Electronic Semiconductor and Logic Symbols	60p
BP28	Resistor Selection Handbook (International Edition)	60p
BP29	Major Solid State Audio Hi-Fi Construction Projects	85p
BP32	How to Build Your Own Metal and Treasure Locators	1.35p
BP33	Electronic Calculator Users Handbook	1.25p
BP34	Practical Repair and Renovation of Colour TVs	1.25p
BP35	Handbook of IC Audio Preamplifier & Power Amplifier Construction	1.25p
BP36	50 Circuits Using Germanium, Silicon and Zener Diodes	75p
BP37	50 Projects Using Relays, SCR's and TRIACs	1.25p
BP38	Fun and Games with your Electronic Calculator	75p
BP39	50 (FET) Field Effect Transistor Projects	1.50p
BP40	Digital IC Equivalents and Pin Connections	2.50p
BP41	Linear IC Equivalents and Pin Connections	2.75p
BP42	50 Simple L.E.D. Circuits	95p
BP43	How to Make Walkie-Talkies	1.50p
BP44	IC555 Projects	1.75p
BP45	Projects in Opto-Electronics	1.25p
BP46	Radio Circuits Using IC's	1.35p
BP47	Mobile Discotheque Handbook	1.35p
BP48	Electronic Projects for Beginners	1.35p
BP49	Popular Electronic Projects	1.45p
BP50	IC LM3900 Projects	1.35p
BP51	Electronic Music and Creative Tape Recording	1.25p
BP52	Long Distance Television Reception (TV–DX) for the Enthusiast	1.95p
BP53	Practical Electronic Calculations and Formulae	2.25p
BP54	Your Electronic Calculator and Your Money	1.35p
BP55	Radio Stations Guide	1.75p
BP56	Electronic Security Devices	1.45p
BP57	How to Build Your Own Solid State Oscilloscope	1.50p
BP58	50 Circuits Using 7400 Series IC's	1.35p
BP59	Second Book of CMOS IC Projects	1.50p
BP60	Practical Construction of Pre-amps, Tone Controls, Filters & Attn	1.45p
BP61	Beginners Guide to Digital Techniques	95p
BP62	Elements of Electronics – Book 1	2.25p
BP63	Elements of Electronics – Book 2	2.25p
BP64	Elements of Electronics – Book 3	2.25p
BP65	Single IC Projects	1.50p
BP66	Beginners Guide to Microprocessors and Computing	1.75p
BP67	Counter Driver and Numeral Display Projects	1.75p
BP68	Choosing and Using Your Hi-Fi	1.65p
BP69	Electronic Games	1.75p
BP70	Transistor Radio Fault-Finding Chart	50p
BP71	Electronic Household Projects	1.75p
BP72	A Microprocessor Primer	1.75p
BP73	Remote Control Projects	1.95p
BP74	Electronic Music Projects	1.75p
BP75	Electronic Test Equipment Construction	1.75p
BP76	Power Supply Projects	1.75p
BP77	Elements of Electronics – Book 4	2.95p
BP78	Practical Computer Experiments	1.75p
BP79	Radio Control for Beginners	1.75p
BP80	Popular Electronic Circuits – Book 1	1.95p
BP81	Electronic Synthesiser Projects	1.75p
BP82	Electronic Projects Using Solar-Cells	1.95p
BP83	VMOS Projects	1.95p
BP84	Digital IC Projects	1.95p
BP85	International Transistor Equivalents Guide	2.95p
BP86	An Introduction to Basic Programming Techniques	1.95p
BP87	Simple L.E.D. Circuits – Book 2	1.50p
BP88	How to Use Op-Amps	2.25p
BP89	Elements of Electronics – Book 5	2.95p